THE FUTURE COMPUTED
AI & Manufacturing

By Greg Shaw

Foreword by Çağlayan Arkan

Published by Microsoft Corporation
Redmond, Washington U.S.A.
2019

First published 2019 by Microsoft Corporation
One Microsoft Way
Redmond, Washington 98052

© 2019 Microsoft. All rights reserved

TABLE OF CONTENTS

01	**FOREWORD**
11	**EXECUTIVE SUMMARY**
19	**CHAPTER 1: THE FUTURE OF AI AND MANUFACTURING**

   Boundless Collaboration: Digital Transformation Means Cultural Transformation

   Seizing the Moment: Capabilities as a Competitive Advantage

   Leveraging the Data Estate: Turning Insights into Action

47	**CHAPTER 2: AI AND THE MANUFACTURING WORKFORCE**

   A New Relationship with Technology: Creating a Safer and More Satisfying Workplace

   The Talent Supply Chain: A Shared Agenda for Skills and Employability

   Working Together

69	**CHAPTER 3: FOSTERING RESPONSIBLE INNOVATION**

   Ethical AI in Manufacturing

   New Rules for New Technology

   Wheres to from Here

109	**CHAPTER 4: THE WAY FORWARD**

   AI Maturity and Progressing Your Journey

   A Path to Innovation

117	**CONCLUSION**

FOREWORD

It has been a long and eventful journey for me from my time as a point guard in the Turkish professional basketball league in the 1980s to my role leading Microsoft Corp.'s Manufacturing & Resources Industry Group today. And while I have traded my sneakers for a computer, the experience of what teamwork can achieve has never left me. Today I am part of a different team—one committed to helping our manufacturing customers adopt, embrace, and lead the digital transformation of their organizations in an era of AI.

The times could not be more exciting for Microsoft and the manufacturing industry, especially when you think of the opportunities that AI will present in defining both *what* manufacturers create and *how* they create it.

This excitement is not surprising given that manufacturing is often a relatively early adopter of new technologies. As one of the first industries transformed by the Industrial Revolution in the 18th century, manufacturing has already established itself as a pioneer in the Fourth Industrial Revolution in this century.

Today manufacturing has a strong association with AI, especially the use of automation. But the automation that we have typically associated with manufacturing in years past was often production lines, heavy machinery and, more recently, industrial robots. These robots were expensive and bulky machines that were mostly used in large factories and out of reach for many small and medium-sized manufacturers because of limited budgets or factory floor space. In recent years, however, thanks to advances in AI technology, these robots have become smaller and far cheaper and are leading a very different interaction and relationship between humans and technology on the factory floor.

Indeed, what has emerged is a whole new branch of robotics, developing what are called "collaborative robots" that are working safely alongside people.

These "cobots" are replacing many of the traditional robots kept in cages or behind safety screens and can be programmed to perform a wide variety of tasks, making them a far more attractive option for manufacturing companies of all sizes and for their workers in all manner of scenarios.

Yet, despite their impact on making manufacturing more efficient and safer, and reducing the number of tedious and repetitive roles, automation and robots have been getting their fair share of criticism. Some critics speak about the need to have "robot-proof jobs"[1] or to tax them like employees if they take the job from a human.[2] But from our experience, and the experience of many of our customers, the reality appears more beneficial than bleak.

Returning to the theme of sports, technology is enhancing human and team performance, and even creating more jobs. For example, professional teams around the world have analytics departments that employ data scientists to analyze and support players' performance. Similarly, new roles are being created in manufacturing to develop new technologies and leverage AI. And this comes amid an existing job shortage in manufacturing.

The Manufacturing Institute and Deloitte estimates that the United States alone may have as many as 2.4 million manufacturing jobs to fill between now and 2028, with around 500,000 unfilled jobs in the sector today.[3] And there continues to be a shortage within the manufacturing workforce, with skilled trades and manufacturing jobs ranked among the hardest roles to fill for the past ten years according to the 2018 ManpowerGroup global Talent Shortage Survey.[4] What is more, as the second wave of the Baby Boomer generation moves into retirement, this trend will only worsen.

It's hard to be certain about the impact of AI on manufacturing. And while automation might displace some jobs and some roles, there are more jobs and opportunities being created as automation and AI helps drive a manufacturing resurgence in many regions. Indeed, PwC forecasts that AI will contribute to a 14 percent increase in global gross domestic product by 2030, or around US$15.7 trillion, as the technology increases productivity, product quality, and consumption.[5]

Other research suggests that AI could double annual economic growth rates in mature economies by changing the nature of work and creating a new relationship between humans and machines.[6] In Germany alone, one government-sponsored study predicts that between 2018–2023, AI will add €32 billion to the country's manufacturing output: This equates to about one-third of the entire growth expected in this sector over the same period.[7]

The future of manufacturing and manufacturers' ability to improve key areas, such as worker safety and sustainability in directly dependent on AI and technology. We hear that firsthand from customers and see the value they are already delivering.

Similar to a coach, my job today is not simply about helping customers with their transformation to an AI company; it is also about releasing this potential of the teams that make up our customers and the broader economy. That is what inspires me and our entire team every day.

Microsoft's approach to help this transformation is to start with the business outcome our customers want. Only then do we identify the necessary insights from their own data estate and what models will help realize the transformation impact they seek.

We want customers to be agile and nimble by breaking down data silos within their organizations and creating a cognitive supply chain where AI can make 30 to 40 percent of decisions, releasing time to do the things that humans do best.

Microsoft has long been a technology partner in manufacturing, and, over time, we have learned how important it is to bring customers into our roadmap planning.

Today our engineers share their roadmaps with manufacturers like Komatsu, Rockwell, John Deere, Airbus, and many others to help shape and reshape priorities.

Beyond this roadmap sharing, we are also co-creating with customers. For example, Microsoft's Power BI was shaped, in part, through work with airplane engine manufacturers like Rolls Royce, and our work with industrial conglomerate thyssenkrupp has helped design our Internet of Things (IoT) priorities.

But we are not just a technology partner with our manufacturing customers—we are a manufacturer in our own right. For over 30 years we have been producing peripherals like keyboards and mice, gaming consoles, personal computers, and mixed reality headsets along with other related products.

We understand the complexities of a global supply chain and the opportunities of applying our own technology solutions to our own manufacturing operations to help us be more efficient, sustainable, and competitive.

The insights we have drawn from our own and our customers' experiences reveal that we have only scratched the surface of the true impact of AI to the manufacturing sector, its workforce, and society more broadly.

To expand on this understanding, we went out and interviewed many of these manufacturing leaders, including our own operations team, to look beyond the hype of AI. We wanted to discern what is real, what is possible, what is worrying, and what stands in the way.

The Future Computed: AI and Manufacturing picks up where the inaugural volume of Microsoft's *The Future Computed: Artificial Intelligence and its Role in Society* left off. This book is a deeper dive into the challenges and opportunities our small, medium, and large customers face across a broad spectrum of manufacturing businesses. It introduces you to the people we are proud to work alongside of and shares their inspiring stories of the progress of AI.

This is the story of inspirational human achievement, winning teams, and organizations. This is the story of companies who are often dreaming big, sometimes starting small, but always aiming to move fast.

ÇAĞLAYAN ARKAN
Global Lead
Manufacturing & Resources Industry Group
Microsoft Corp.

ÇAĞLAYAN ARKAN

WHAT AI MEANS TO MICROSOFT

When you ask someone what they understand by the term "AI," the response is often framed by how the technology is portrayed in science fiction novels or Hollywood blockbusters. But, as with many technologies, the reality of AI is less sensational and more practical than our imagination is led to believe.

Put simply, AI is a machine's ability to recognize patterns, sounds, images, and words, and to learn and reason over data—to infer—in ways that are similar to what people do. It's a set of technologies that enable computers to understand and interact with the world more naturally and responsively than in the past, when computers could only follow preprogrammed routines.

AI isn't new, and it certainly isn't new to Microsoft. In fact, computer scientists at Microsoft and elsewhere have been working on machine learning and AI technologies for decades, and we are now witnessing the realization of this work in a wave of breakthroughs that promise to open a new era for AI applications in business and society.

This move from the computer lab into mainstream products is thanks to a set of contributing factors: the massive computing power of the cloud; the availability of enormous datasets that can be used to teach AI systems; and the breakthroughs in developing AI algorithms and improving AI methods such as deep learning.

At Microsoft, we have advantages in every one of these areas.

The immense computing power of the Azure cloud offers customers far more than just a place to store data. It allows companies like Rolls Royce to discover actionable insights around fuel usage, predictive maintenance, and stopping unscheduled delays for their aircraft engines. It's also the technology behind Danish brewer Carlsberg experimenting with whether AI can help predict how new beer varieties will taste.

But AI systems are only as good as the data from which they learn. We work with our customers to create AI solutions that will help harness *their data* and create better insights for their organizations.

When we add AI capabilities to existing products, or release new products infused with AI, they are often rooted in discoveries from Microsoft's research labs.

Microsoft researchers have led breakthroughs in fields including image recognition, machine translation, speech recognition, and machine reading comprehension, which uses AI to read, answer, and even ask questions. Those breakthroughs then make their way into products such as Azure Cognitive Services, a set of tools developers can use to add AI capabilities such as image recognition, translation, and visual search into their products, and the bot-building tools in Azure Bot Service.

Taken together, these advantages are enabling Microsoft and our customers to create products and services that use AI to better understand, anticipate, and respond to people's needs.

One example of AI in use is Cortana, Microsoft's intelligent digital assistant. Cortana can do everything from scheduling a meeting for you, to reminding you to follow up on a specific commitment you made in an email. Cortana is able to do things like give you reminders because it's been trained using massive, anonymized datasets to understand why a specific email is important. Cortana will then learn specifically what emails might be important to you based on your behavior and preferences.

But beyond digital assistants, Microsoft's vision is to democratize AI even further. To realize this, we envision a "computing fabric" for everyone, one that blends the intelligent cloud with the intelligent edge—where data is analyzed and aggregated closest to where it is captured in a network. We envisage a world where this cloud and edge computing is coupled with a multi-sense, multi-device experience that seamlessly integrate speech, gestures, and the gaze of our eyes. This is a world where mixed reality—virtual worlds with digital twins—will become the norm.

"We are creating new experiences and enabling digital transformations for every customer," Microsoft CEO Satya Nadella told shareholders at the 2018 annual meeting. "We're leading the field of AI research, achieving human parity with object recognition, speech recognition, machine reading, and language translation. But, most importantly, we are focused on democratizing these AI breakthroughs to help organizations of all sizes gain their own competitive advantage, because of AI."

But AI is not simply about productivity and efficiencies; it is about providing resources and expertise to empower those working to solve humanitarian issues and create a more healthy, sustainable, and accessible world.

One example is from our team at Microsoft Research in the U.K. who are working with a group of oncologists on a project called InnerEye that uses AI to mark the boundaries between healthy tissue and tumors in CT scans. Today this is a time-consuming job that is done by hand—pixel by pixel across in hundreds of layers of scans. Thanks to InnerEye, the job can be accomplished in seconds.

Another powerful example is a tool called Seeing AI, which was developed by Saqib Shaikh, a Microsoft engineer who lost his sight at age seven. This powerful mobile app reads signs and documents, identifies currency and products, recognizes friends, and interprets peoples' facial expressions so it can describe what it sees to the user through an earpiece. Seeing AI allows people with low vision or blindness to navigate the world with greater independence.

Finally, with the world's population expected to grow by nearly 2.5 billion people over the next quarter century, AI-driven farming offers hope that we'll be able to grow enough food to meet the increased demand. Our FarmBeats project is using the power of the cloud and machine learning to improve agricultural yields, lower costs, and reduce the environmental impact of farming.

At Microsoft, we believe that these stories provide just a hint of the unprecedented opportunities that AI offers to address pressing challenges, drive progress, and transform how businesses operate and people work.

EXECUTIVE SUMMARY

When Bill Gates and Paul Allen founded Microsoft four decades ago, their goal was to democratize the benefits of software for computing—then largely restricted to enormous, expensive mainframes—to everyone. Their vision was a computer on every desk and in every home. Today, Microsoft is aiming to do something similar with AI—to make AI development and its benefits available to everyone.

Our AI systems are designed to empower people and to augment their capabilities, and we're committed to making sure our AI tools and technologies earn their trust.

To help advance the conversation on how AI can empower people while fostering trust in the technology, in January 2018 Microsoft published *The Future Computed: Artificial Intelligence and its Role in Society*. The book represents our perspective on where AI technology is going and the implications for society in a future where the partnership between computers and humans will be more pronounced than ever before.

Microsoft President, Brad Smith, and Executive Vice President of Microsoft AI and Research Group, Harry Shum, wrote in the book's foreword that AI will enable breakthrough advances in every industry while raising complex questions and concerns. They sought to answer some fundamental questions, such as: How do we ensure that AI is designed and used responsibly? How do we establish ethical principles to protect people? How should we govern its use? And how will AI impact employment and jobs? Ultimately, the question they posed is not what computers can do, but what computers should do?

In this companion book—the first of a series to explore AI, the future of the workforce, ethics, and policies related to individual industries—

we delve into manufacturing, an enormous global ecosystem and the backbone of our human economies. It's a sector that throughout history has embraced new technologies while shaping the nature of work and creating enormous economic and societal impact.

When we first conceived this book, we had three audiences in mind: manufacturers who are curious about the AI journey of their colleagues; policymakers and regulators who are examining AI's societal implications; and our commercial partners who want to stay abreast of both the technological breakthroughs and the policy possibilities.

In scores of interviews with manufacturers, workforce experts, unions, and policymakers, we heard time and time again that manufacturers and workers alike are embracing this AI journey. In the pages that follow, we hear customers of Microsoft describe that journey, including their views that this technology has profound implications for their business models, their workforce, and their responsibilities as ethical organizations.

We hear from labor organizations that they are encouraged about how AI can improve worker safety and job satisfaction, but they are pragmatic about the pace of technology change. They recognize the need to be planning for the impact on workers, especially by building adaptable skills to meet the opportunities and be more resilient to change.
We hear from policy makers that they are excited by AI's potential to help drive jobs and economic growth, but they are anxious to have the right policy environment to help realize the technology's potential in a responsible way.

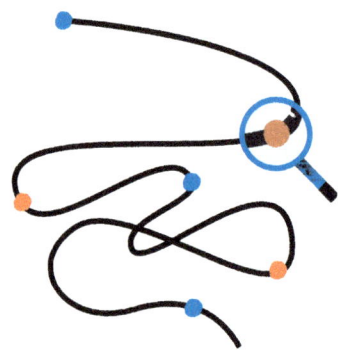

Listening to our customers and these stakeholders, we discovered six themes:

1. **Manufacturers are already seizing the AI opportunity.**
   AI is not simply about productivity. It's also about helping reinvent organizations. AI is about workplace safety and health, predictive maintenance, process efficiencies, intelligent supply chains, up-time, higher value, and higher-quality products. These are the drivers of AI design and implementation for the manufacturers we spoke with. In addition, as a global manufacturer, Microsoft is using AI to drive its own transformation, and we are sharing those insights with our customers.

2. **Central to digital transformation is cultural transformation.**
   Strong leadership combined with engaging workers at all levels in the process is essential. In order to optimize AI's value, the entire organization must work together to embrace change, break down silos, and create a seamless information supply chain inside companies and leverage their full data estate.

3. **Those closest to the workforce, the managers and leaders inside manufacturing operations, are often the most sensitive to AI's impact on their workforce.** Their focus is to create a better company and more opportunity, including a safer work environment, fewer repetitive and unsatisfying jobs, and increasing productivity. They put their people first and so are eager to adopt technologies that have positive impact on workers.

4. **There will be disruption and dislocation, and we need a new pipeline of talent.** While there is positive sentiment about the opportunities for the workforce, manufacturers are also concerned about talent shortages, short-term disruption, and attracting the next generation of bright students. Jobs in manufacturing will require new skills and new capabilities, and this necessitates a new partnership for skills and workforce development with technology providers, industry, government, learning institutions, and labor organizations.

5. **Next-generation policies and laws are needed for next-generation technologies.** Our customers are already versed in many of the ethical issues that are associated with AI—security, safety, reliability—but are looking for more guidance on how to use AI ethically and responsibly. As manufacturers, they are using AI in their production processes and infusing AI into their products, and they are therefore looking for guidelines that will help anticipate potential issues and ensure responsible innovation. Similarly, regulators are eager to remove technology hurdles and encourage the adoption of AI technologies that will promote worker safety, create more jobs, and help national competitiveness.

6. **AI is a journey, different for everyone.** While many of our leading customers have embraced AI, there are many who are just beginning their journey. Microsoft has developed a guide to help them on their AI journey and show them how to best leverage their data estate. Implementation ranges from improving reporting through data analytics and business intelligence to "deep" AI capabilities through machine learning applications.

But what do we take from these themes and how do we move forward with these insights?

First, it's becoming increasingly clear that society needs a long-term and multi-stakeholder approach to realize industry transformation potential through AI. As the lines between industry policy and technology policy continue to blur, coupled with the progress of manufacturing companies becoming digital companies, the need for our customers to engage in the development of the digital policy agenda is more important than ever.

So, just as Microsoft is seeking to better understand the industry policy landscape, we are also encouraging our customers to be more engaged in the development of digital policies. We want to bring together leaders across private and public sectors, civil society, employees, and labor organizations to co-create the right policy and ethical frameworks where industry opportunity and human potential can both flourish.

EXECUTIVE SUMMARY

Second, all of us need to prepare for the impact on the workforce and build a supply chain of talent to help new workers coming into the workforce acquire the new skills they will need, develop workers who will have to transition to new jobs with the same employer, and support those whose roles will be eliminated and will need new jobs elsewhere in the economy.

But, finally, what this really means is we must keep engaging. Just as we outlined in the original Future Computed, we're all going to need to spend more time talking with, listening to, and learning from each other. No single company has all the answers, and this book doesn't seek to present all the issues that an industry as broad and as diverse as manufacturing is facing. This book is just part of our ongoing engagement and dialogue to surface issues and build coalitions of interest to address them. We continue to look forward to working with people in all walks of life and every sector for building a foundation for AI that transforms not just businesses, but transforms lives.

THE AI POTENTIAL: THE DATA

THIS INFOGRAPHIC IS BASED ON MICROSOFT ANALYSIS OF THIRD-PARTY DATA. SOURCES INCLUDE:

(PWC)

1. Manufacturing the Future: Artificial intelligence will fuel the next wave of growth for industrial equipment company. Accenture 2019 https://www.accenture.com/t20180327T080053Z__w__/us-en/_acnmedia/PDF-74/Accenture-Pov-Manufacturing-Digital-Final.pdf#zoom=50

2. McKinsey – Digital Manufacturing Capturing Sustainable impact at scale (June 2017)

3. IDC FutureScape: Worldwide Operations Technology 2017 Predictions Jan 2017 Doc # US42261017 Web Conference By: Lorenzo Veronesi, Marc Van Herreweghe.

4. https://futureoflife.org/national-international-ai-strategies/?cn-reloaded=1

5. The Economist Intelligence Unit, Intelligent Economies: AI's transformation of industries and society, (July 2018) https://eiuperspectives.economist.com/technology-innovation/intelligent-economies-ais-transformation-industries-and-society?utm_source=Organic%20Social&utm_medium=Twitter&utm_campaign=Microsoft%20-%20Intelligent%20Economies&utm_content=Briefing%20Paper

6. World Economic Forum, Future of Jobs report 2018 (September 2018) http://reports.weforum.org/future-of-jobs-2018/

7. American Council for an Energy Efficient Economy, How Smart Buildings Save Energy, (Nov 2015) https://www.buildings.com/article-details/articleid/19537/title/how-smart-buildings-save-energy

8. Manufacturing the Future: Artificial intelligence will fuel the next wave of growth for industrial equipment company. Accenture 2019 https://www.accenture.com/t20180327T080053Z__w__/us-en/_acnmedia/PDF-74/Accenture-Pov-Manufacturing-Digital-Final.pdf#zoom=50

9. Food and Agriculture Organization, How to Feed the World in 2050 (2015), www.fao.org/fileadmin/templates/wsfs/docs/expert_paper/How_to_Feed_the_World_in_2050.pdf

10. OECD, Water Outlook to 2050: The OECD calls for early and strategic action, (May 2012) www.globalwaterforum.org/2012/05/21/water-outlook-to-2050-the-oecd-calls-for-early-and-strategic-action/

11. World Economic Forum, Future of Jobs report 2018 (September 2018) http://reports.weforum.org/future-of-jobs-2018/

12. IDG Communications, Top 5 cybersecurity facts, figures and statistics for 2018 (January 2018) https://www.csoonline.com/article/3153707/security/top-5-cybersecurity-facts-figures-and-statistics.html

13. IDG Communications, Top 5 cybersecurity facts, figures and statistics for 2018 (January 2018) https://www.csoonline.com/article/3153707/security/top-5-cybersecurity-facts-figures-and-statistics.html

14. Cybersecurity Tech Accord, https://cybertechaccord.org/about/

CHAPTER 1

THE FUTURE OF AI AND MANUFACTURING

THE FUTURE **OF AI AND MANUFACTURING**

Whether standing on a plant floor or in a cavernous warehouse, manufacturers light up when talking about efficiency, productivity, and performance. They lower their eyes and sigh when they speak of downtime and legacy machines. But the potential of smart factories and their own individual roles in leading a small part of a much larger AI revolution is evident in every conversation with manufacturers across the world.

Very often those we interviewed are engineers—or entrepreneurs with engineering inclinations—who rose through the ranks. They didn't set out in life to lead an AI revolution, but their deep knowledge of the inner workings of their manufacturing processes thrust them onto the stage of next-generation technologies. They are a bridge between the old and the new.

The following chapter is about sharing their stories and what they can teach us about what Microsoft CEO Satya Nadella calls "tech intensity." This is where the potential for companies and countries to jump-start their growth by not just adopting technology, but by building their own technology and skills too. One way to think about tech intensity is creating the right environment for the fusion of cultural mindsets and business processes—a fusion that rewards the development and propagation of digital capabilities to create boundless collaboration and results in new insights and predictions, automated workflows, and intelligent services.

Understanding tech intensity is best illustrated through the stories of our customers—stories that span meaningful results such as collaboration, competitive advantage, and creating better data supply chains.

BOUNDLESS COLLABORATION: DIGITAL TRANSFORMATION MEANS CULTURAL TRANSFORMATION

The Ups and Downs of AI

The sound of a phone ringing in the middle of a weekend night awakened Patrick Bass from his deep sleep. As a young elevator technician, Patrick was not completely surprised because it had fallen to him to be on call if a customer needed help. What did surprise him was the voice on the other end.

It was an elderly woman who was trapped in her home elevator. She was understandably upset and gave him instructions for how to get into her house. He threw on some clothes and sped over immediately. The situation he discovered was troubling. The woman's robe had been caught in the elevator system, and she was only partially clothed. The reason she was upset had only a little to do with her own predicament; she was mostly upset that she had to disturb Patrick in the middle of the night.

"Her sincerity is unforgettable," he said. "It helped me work through the uncomfortable situation."

Patrick, who today is the CEO of thyssenkrupp North America, is indicative of so many manufacturers we spoke with—brilliant engineers pioneering new technologies and sensitive leaders with profound emotional intelligence.

Born on a dairy farm near Burlington, Wisconsin, his mother had been confined to a wheelchair since age 13. "She wasn't supposed to have kids, so I am a mom's boy."

He was able to care so perceptively for the woman caught in the elevator because he had been caring for his mother for as long as he could remember. Helping people was a natural part of his life.

Patrick loved systems and understanding how things work. He knew milking cows was not his long-term future, and his mother and grandmother were intent on him going to college. He got an engineering degree and always dreamed of designing cars.

His first job was working with an original equipment manufacturer to design disability products for automobiles. In time, Patrick moved on to a company called Mobility Unlimited. As fate would have it, his family also had an electronic business which had enabled him to get an electrician's certificate. The elevator mechanics at his new employer needed a certified electrician to sign the license, and so he became an elevator helper, simply handing tools to the mechanic. He fell in love with the job and began to service the products Mobility Unlimited didn't want to service. He once worked on a service elevator dating back to 1917 that employed sheets of leather with water hydraulics to pull cables from a restaurant below ground to the sidewalk above. There was no manual to read so he had to figure things out on his own.

He went from helper to full lead mechanic in a year. Today Patrick has spent 26 years in the elevator industry, working on all facets from designing and manufacturing to servicing and selling elevators.

He started at thyssenkrupp as Product Engineer "Number 1" and progressed up the chain of command from there. He had gone from a small elevator company to a very big one. Eventually he was assigned to the research group and became fascinated with the data generated by elevators. He pushed himself and the team to understand how they could extract data from elevators to make predictions about maintenance and safety.

"We failed miserably," he said. "It left a tarnish, but we regrouped and worked through it."

Years later, he sat in a corporate meeting and learned about a "black box" project in the field that involved streaming elevator data to a mechanic. The presentation focused on the device, not the data, the cloud, or AI. He spoke up.

THE FUTURE
OF AI AND
MANUFACTURING

"I've been here before and you are focused on [the] wrong things," he told them.

An executive called for a break to conference with Patrick, and when the meeting reassembled, Patrick was put in charge of the project. But that wasn't all.

"They can either reshape it or kill it," the executive declared.

Fate again struck. Just as Patrick was reimagining the project, he went to see the new CEO of Microsoft, Satya Nadella. Satya had been in the new role only a few months. They met and realized they shared a vision about what data and AI could be.

Patrick said, "Satya looked at us and said, 'Let's do this.' We will do a proof of concept together, but we're going to do this in three months, and we need you to keep up."

Like the proverbial phoenix that rises from its own ashes, thyssenkrupp's Max—Maximum Uptime, All the Time—was born. It was the elevator industry's first real-time, cloud-based, predictive maintenance solution. With more than 12 million elevators moving one billion people every day, uptime is important to any elevator, whether it's a thyssenkrupp elevator or not. But digital transformation had come to thyssenkrupp quickly—and digital transformation means data is available to everyone all the time, in real time, anytime, and on any device for the purpose of empowering people and business.

"One source, one truth," Patrick recalled. "Everything is a system. People systems are the most dynamic and most rewarding. My job as a leader is to help empower people."

Since the initial success, thyssenkrupp has taken this cultural transformation to help other customers. For instance, one auto manufacturing customer of theirs was struggling with one specific operation on the assembly line—sealing a door. It was a costly, time-laden job, but it was also a job that AI could help improve. thyssenkrupp realized the customer was not scaling data beyond a single factory in order to get bigger impact across the company. The factory simply said "no" when asked for its data. "Sure, you are corporate but you're not taking my secret sauce," the factory manager said.

Over time, though, the manufacturer was able to get everyone on board to share data. Stakeholders came to understand the data was not going to be used against them. Instead, they shared data to build a digital twin—a virtual assembly line—to come up with a better solution. They now have a brand-new processing system that takes not just one piece of data, but all the data necessary to design a new system. Patrick and team were able to go to the customer and present a new solution. Overnight, they had hundreds of millions of dollars in new revenue from top-tier automotive producers.

"Everything is a system. People systems are the most dynamic and most rewarding."

Silos with a View

Thyssenkrupp's story of cultural transformation is not isolated. The collaborative work Patrick and his team led to bring new products and services to life is mirrored at another of our customers, Jabil.

Jabil is a global manufacturing solutions provider with more than 200,000 people across 100 facilities in 29 countries. One of those facilities is in Tampa Bay on Florida's west coast, and this is where we meet Matt Behringer, Jabil's VP of Operational Technology.

Matt bristles when you ask if AI can tear down silos to make the business more connected and agile. He will correct you. In fact, he argues, every business needs some silos to ensure functional depth. Those silos, however, need windows to provide adjacent businesses and services with the visibility they need to collaborate and optimize for customer needs.

Matt is leading Jabil's "Factory of the Future" initiative. The vision is to create an autonomous, adaptable factory.

"We need things connected and interacting so that when change occurs, the technology understands what's going on and can adapt to that change," Matt said. "Equipment, people, and technology—all of it."

To accomplish this, they've moved from organizing by traditional internal functions to cross-functional workstreams. Automation across the entire organization is the goal. A connected system will correct itself before there is a problem. Is the system drifting or are there human issues? In the past, sensors and cameras were not far enough along, but today they are.

"Don't make a defect, fix it now," Matt said. "Do it automatically rather than using charts to diagnose later."

Even in an era of automation and AI, Jabil sees humans continuing to be central to the equation. And while people may be repurposed and retrained, giving humans real-time feedback to make improvements is essential.

Matt should know. Born in Illinois on a dairy farm, he moved as a teenager to Florida and after college became a general contractor on boats. This is where he met one of the founders of Jabil who noticed his technical and entrepreneurial skills and quickly hired him. Starting in production in a newly constructed building, Matt moved up quickly in the manufacturing operation. At the time, Matt said he couldn't even spell IT; but Jabil asked him to learn information technologies and over time his math and engineering skills absorbed that as well. He became fascinated with how the company could empower people to do their jobs better. Today he is the vice president in charge of operational technology, including Jabil's Factory of the Future initiative.

"We are not taking away silos, but we are putting windows in them so we can compound the value proposition—aim everyone in the same direction. We need silos but we need them harmonized."

The experiences of thyssenkrupp and Jabil tell us that cultural changes are as important as the AI technologies in helping empower employees and make data-driven decisions. These organizations are focused on adopting a data culture and continuing to utilize prioritized strategic AI initiatives to disrupt their industry and create new business models and streamline operational processes. Next, we introduce you to some equally inspiring organizations that are harnessing AI to carve out competitive advantage.

SEIZING THE MOMENT: CAPABILITIES AS A COMPETITIVE ADVANTAGE

Paper Planes

Manuel Torres was born and raised in the Spanish city of Murcia. An entrepreneur and self-taught engineer, Torres has made the long (and unlikely) journey in manufacturing from pulp and paper, to aerospace, to AI. Early in his career, during the 1973 oil crisis, prices for raw materials like paper increased nearly 100 percent and scrap became an important engineering challenge.

The paper company Manuel worked for, like the industry itself, was struggling with outdated, inefficient processes. There was a lot of scrap, which meant a lot of waste and lost profit.

Torres had an idea for automating the splicing of one massive roll of paper onto another by maintaining proper tension and melding them together, reducing the scrap to zero.

But his employer refused to invest in his idea. Not willing to take no for an answer, he founded his own company, MTorres in Pamplona, Spain, the city famous for the annual running of the bulls. Manuel's young company soon became a leader in paper converting.

At the very beginning, MTorres started selling its splicers to S&S Corrugated Paper Machinery, an American company that built corrugated paper lines. Over time, major brands like Weyerhaeuser, Georgia Pacific, and Kimberley Clark became customers—each eager for technological innovation. A decade later, the aerospace industry came knocking when it needed an innovative way to weave composite materials, like reels of carbon fiber, together for airplane wings, fuselages,

and tails. The move into aerospace was propitious. Not only would the industry need MTorres' capabilities in the near term, they would also be needed for cutting-edge new endeavors.

The Seattle Times reported in early 2019 that Boeing is investing in Nevada-based Aerion, which is developing an advanced supersonic business jet.[8] The 12-passenger AS2 will be built mostly from carbon fiber composite and is designed to fly over ocean routes at speeds up to Mach 1.4, or approximately 1,000 miles per hour. For MTorres, it turns out the same technological insights required to maintain tension in the splicing of paper also apply for more complex materials like carbon fiber. Suddenly Boeing, Airbus, Fokker, Embraer, and many other aerospace manufacturers needed the company's help. Today, there is hardly a jet in the air that has not been touched by MTorres.

The company now holds 140 patents and has plants in 10 locations across four continents, including one just down the road from Boeing outside Seattle. The Spanish manufacturing equipment and automation specialist is now a cherished partner throughout both aerospace and paper.

But the march of progress has hardly slowed. Like so many of the manufacturers we spoke with, MTorres began to realize that not only did it have a competitive advantage due to its experience and skills, it also had a comparative advantage because of its data.

That data coupled with an algorithm and compute power—a combination central to advanced machine learning and AI—is leading MTorres to breakthroughs never before imagined. The ability to flawlessly laminate carbon fiber onto an airplane wing at 60 meters per minute means not only greater efficiency and productivity but also greater accuracy and fewer mistakes. With AI, the company can also inspect its work in real time. Never satisfied, they plan to increase that speed through AI-based real-time inspection.

MTorres now is using advanced machine learning and AI in its Automatic Fibre Placement machines to optimize for scrap reduction. This is achieved via a drill that uses visual recognition to perfect tens of thousands of holes per part, and an automated inspection module that scans the surface of carbon fiber in real time looking for defects with a high resolution camera and a laser. The defect analysis is done using deep learning algorithms in the preprocessing and classification steps. In aerospace, as the saying goes, failure is never an option.

Swarm Mentality

In locations around the world, Toyota Material Handling—the division of Toyota Industries Corporation focused on logistics—is using AI to reinvent the way it does business.

Axel Wahle, a genial leader at Toyota Material Handling, has spent more than 30 years in the business. Rising through Germany's much-respected apprenticeship programs for both engineering and commerce, it's Axel's job to bring the customer voice into everything the company does.

Toyota Material Handling is like many businesses around the globe that are confronting the need to store, move, and deliver components and finished product to help meet consumer demand. Their enormous warehouses are fields of shelves, stacked floor to ceiling and loaded with scores of car parts in every shape and size. It's the job of people like Axel to help manage the material flow for every SKU variation using autonomous pallet drones and forklifts that snake their way across warehouses to retrieve goods needed—whether for assembly in manufacturing applications, or for consignments within the retail sector.

Not long ago, Axel presented a new idea involving AI that would improve productivity and accuracy. It's called "swarm" and represents a revolutionary new approach to what the company calls "automated handling." Small robots swarm through the facility and use AI with visual recognition to navigate. They carry loads to the correct location for consolidation and onward transportation.

Toyota Material Handing uses Microsoft's AirSim, which creates realistic digital twin environments to replicate vehicle dynamics and test how autonomous vehicles using AI can operate safely in the open world. Further, the company uses Microsoft Bonsai, a novel approach using machine teaching that analyses the images taken of an environment to identify components such as racks, pallets, and other infrastructure. Then the autonomous vehicle training takes place inside the AirSim simulated environment.

The end vision is that Toyota Material Handing's AI system can examine the image of a shelf full of products, identify bar codes, and get the right product to the right place. The vehicle is constantly learning and then sharing what it's learning with other vehicles.

Axel said, "The swarm is for coordinated distribution, lean logistics, and continuous flow—the right truck, the right task, on time, every time by utilizing millions of calculations per second."

A human is ultimately responsible for validating that all the parts are correct. Given the diversity of SKUs in which variations are so slight, Toyota Material Handling entrusts final approval to human eyes and experience. The approach made sense to engineers, but the company insisted that customers be consulted for feedback. More than 1,800 customers provided feedback on the swarm approach and, after careful examination, Toyota Material Handling agreed that they should pursue it because the ROI for humans is so great and the accuracy is so high for the end consumer.

What the two examples—MTorres and Toyota Material Handing—demonstrate is thanks to the digitization of assets and the infusion of AI to automate processes, these organizations are applying AI not only to their current processes, but they are creating additional value and carving out competitive advantage.

The next customer stories will share how manufacturers are thinking about data in the same way as they consider other critical parts of their physical supply chain.

TURNING INTELLIGENCE INTO ACTION: CREATING A DATA SUPPLY CHAIN

It's AI, but Not as You Know It

In Detroit—a city famed for its role in the Second Industrial Revolution—we visit a customer helping to define the Fourth Industrial Revolution.

Auto parts giant ZF Group is pioneering data algorithms and technology to make their production line more reliable and sustainable. In other words, detecting and predicting mechanical failures on the manufacturing line and conserving energy. But to create the right culture of transformation, Georg Gabelmann, ZF's Data Science Manager, says, "Don't call it AI. For me, AI is a buzzword."

> **A company needs to continue its path to capture quality data and to develop the human skills to improve machine learning and eventually build true AI.**

Georg is an industrial engineer with experience at SAP and thyssenkrupp who now leads the IT Innovation Team at ZF where they invent, showcase, and prove a range of new methodologies. Georg says, "True AI is miscommunicated and misunderstood. Applied machine learning is AI, and data science is the key. If someone came to me and said we are doing AI, I would disagree. Our first step is machine learning." Georg believes the company needs to continue its path to capture quality data and to develop the human skills to improve machine learning and eventually build true AI. And they are making progress. For example, a situation arose where a crucial tool on one of ZF's lines was continually breaking and no one could figure out why. Working with Microsoft, they trained a model to diagnose the problem and, over time, by using more and more sensor and maintenance data, they improved the model and eventually were able to better predict the expensive tool failures.

Likewise, they've trained their data models to examine energy consumption. The company pays a premium for power when it surpasses a certain level. In the past ZF had to fly blind, but today the company uses an intelligent assistant based on reinforcement learning to better predict energy consumption. An area of machine learning, reinforcement learning is defined as being "concerned with how software agents ought to take actions in an environment so as to maximize some notion of cumulative reward." Georg explains that his machine learning system can recommend when to shut down an air conditioning unit or to postpone a test run until later.

Georg's colleague, Robert Copelan, is based in Atlanta and has spent 33 years in the automotive industry. As he listens to Georg talk about where ZF is today, he cautions that the company is still very much in a proof-of-concept phase. Progress is not yet enterprise-wide. Asked if he had advice for other manufacturers, Robert replied, "Think big but start small. You have to think about what it means to have it adopted across the entire enterprise. Showcase and create awareness that can prove the value and get people on board."

"Think big but start small."

He pointed out that some workers will fear machine learning and AI. ZF has worker councils designed to explain these advanced technologies. "AI will work in conjunction with us, getting rid of repetitive work," Robert noted. "We strongly recommend not shutting down new tech initiatives, instead trying to understand and take away the fear."

Workers are already embracing the idea that having algorithms with them when they need them is important. ZF's "XReality" is a combination of augmented and mixed reality in which wearable devices like Microsoft's HoloLens can help workers identify machines and parts, access data, and help a worker or floor supervisor solve a problem.

ZF is helping to make workers more autonomous even as they help car manufacturers build their autonomous and electric vehicles of the future.

Milking the AI Potential

While cars might be driving themselves, dairy cows can't milk themselves, but AI is helping one of our customers take a step closer. The urgent phone call that a food manufacturer makes four seconds after a failure on the dairy line is very different than the preventive call that a manufacturer receives four weeks before a failure.

The call four seconds after a failure is often addled by panic because 50,000 liters of milk are at risk. The call after the prediction of a failure, four weeks before it happens, is somewhat less frantic.

Let's meet someone who is trying to avoid the panic calls by using AI to makebetter predictions in the food manufacturing sector.

Johan Nilsson oversees digital for Tetra Pak, a leading liquid food packaging manufacturer based in Sweden. Tetra Pak does aseptic packaging and processing for major food and beverage brands like Nestle, Tropicana, Coke, and Pepsi. Nilsson tells the story of how AI is helping Tetra Pak prevent failures. Not long ago, a major dairy producer located just outside of Buenos Aires, Argentina, received a call predicting a future failure. By connecting packaging lines to the Microsoft Azure Cloud, Tetra Pak can collect operational data from the dairy to help predict maintenance timing. The consequences of not predicting a failure are far-reaching as Argentina has strong import restrictions, so getting a replacement part from nearby Brazil or far-away United States

can take 96 hours to clear customs. Since a typical cow can produce 24 liters of milk every day, the loss of milk per cow, multiplied by a herd of hundreds, can be the difference between profit and loss. You can't simply switch the cow on and off during a manufacturing failure.

"We can plan with parts so there is no fuss. No loss of production time," Johan boasts.

For many customers, Tetra Pak is the largest supplier they have. They rely on Tetra Pak every day of every month. Having a high-quality relationship is essential. Tetra Pak produces 600 data points per piece of equipment with high frequency—too much data for any human to analyze in real time. The company increasingly is a user of every Azure feature, and engineers are testing blockchain on the Microsoft cloud, which is tamper-proof.

Liquid food factories are held to very strict quality management standards to secure food safety. Those not relying on the cloud and AI cognitive services might be using paper or multiple spreadsheets, which is labor-intensive and creates risk.

The Connected Welder

Predictive maintenance is driving machine learning and AI adoption throughout the manufacturing sector, but so too are operational efficiencies and improving productivity.

Colfax, based in Annapolis Junction, Maryland, provides a case in point. The company, known for its expertise in welding and pumps, has grown from a specialty fluid-handling company to a multi-platform diversified industrial company serving hundreds of global industries.

Colfax is several years into its "Data Driven Advantage" digital strategy, in which it is deploying a global common footprint that will enable the company and its subsidiaries to take advantage of data no matter where it is generated. Ryan Cahalane, a controls engineer turned technology executive with a long resumé in manufacturing, is helping to lead the transformation by aligning Colfax's systems across all infrastructures.

"We used to be the nerds at the dance no one wanted to talk with, but now we're the most interesting people at the dance," he laughs.

One example is Colfax's ESAB subsidiary, a leading global provider of fabrication solutions. With men and women operating their welding, cutting, and gas management equipment in almost every industry and application imaginable, it was very difficult to be able to identify when customers were running into issues, determine the patterns leading up to the event, or make recommendations. In response to the challenge, ESAB had the ingenious idea of creating "the connected welder." By connecting every welder to the platform, AI is lighting up the ability to provide valuable insights into things like job quality, efficient use of material, optimizing consumables, and even alerting users to potential supply chain issues. As a result, ESAB can improve job costing and quality for their customers, or help them make better decisions about critical production resources and equipment utilization. With all that data, and with the ability to finally "see" trends in the huge diversity of customer applications, AI will in turn help make recommendations back to ESAB on how to improve designs and their own operations.

"AI can find patterns we didn't know before—weaving together a fabric that extends across the entire value chain," Ryan explains. "As it unfolds, we will get insights at lots of different levels. We can then make better hypothesis and correlations, at a rate of innovation that was never achievable before."

Ryan has spent his life in and around manufacturing. Originally from Akron, Ohio, his father was a manager at Goodyear, which meant living wherever Goodyear had its tire and rubber plants, be they in France, the Congo, or Morocco. After studies at Purdue and Case Western universities, Ryan joined Goodyear himself, but he eventually went to Deloitte because he wanted to see the bigger picture in his chosen field.

Ryan recognizes manufacturing still has a long way to go, but he also is excited by the meaningful and accelerating progress that lies ahead.

"But it isn't all about new products or 'greenfield' operations," he says.

"There's a huge opportunity in the 'brownfield' (existing/legacy). Where it can have a meaningful impact now, right away, is unlocking value from existing assets. And empowering people on the frontline versus just being tools to help data scientists back at corporate." For the near term, AI is really about 'augmented intelligence,' helping the human, more than artificial intelligence.

"AI is really about 'augmented intelligence,' helping the human, more than artificial intelligence."

But most plants have a mix of new and older equipment, and often the expertise needed to operate at full potential is held by just a few people. "People can make or break the quality of the work on a given day," Ryan points out. "An operator of a line or supervisor is a bit like the conductor of a symphony, and their experience and mindset can make the performance just average or really amazing. Many can tell just by the background sound of the operation how things are running, where problems are beginning, and how long they have before it really affects overall performance."

With the changing workforce in manufacturing, including the retiring of experienced baby boomers, much of this experience and capability is walking out the door. While more automation and robotics may help, humans are still critical in most operations. This means that manufacturers need every operator to perform like a concert master.

Just like the background noise example, AI can help identify patterns that less experienced operators might miss, accelerate learning new skills or amplify skills operators already have, and help operators connect with other experts to collaborate on a particular situation. The old way of controlling an operation was static, often not taking advantage of the incredible capacity operations teams must have to improve performance or innovate.

Now, through AI, the operator is empowered with more knowledge about the baseline throughput and quality, and now is provided with recommendations in real time on how to improve performance. In this case, as with so many others that Colfax have found, operators can now use their full potential to evaluate and execute the best possible performance.

As Ryan observes, "People don't wake up and think about digital transformation. In fact, in the real world of production operations, it makes them recoil. If not positioned and focused properly, AI could cause a real rift with people if we're not careful. Augmented intelligence is a more powerful, and timely, message. That is something people can get behind. This will make me a better worker and compete more with the offshore, low-cost worker that threatens to take my job."

From car parts to beverage packaging, from industrial welding to elevators, this diverse range of manufacturers demonstrates the reach and impact of AI. These customers understand the value of data and how essential it is to making their products safer and more reliable. Each manufacturer recognizes their critical role in a larger supply chain and that AI will be an ever-increasing asset to liberate and harness data across that supply chain to meet their needs and the needs of their customers.

Where to from Here?

In story after story, we met individuals who care deeply about the future of their business as well as the future of their workers. We heard from managers and plant workers who understood that digital and cultural transformation work hand in hand. We also witnessed what happens when you remove the barriers around an organization's data estate and convert information into innovation.

And while these stories show us what realizing tech intensity is all about, they also remind us that people are at the heart of this transformation. In the next chapter we look carefully at questions and issues surrounding the human impact and the implications for tomorrow's AI workforce.

AI AND MANUFACTURING: MICROSOFT'S CULTURAL TRANSFORMATION

The time difference between America's Pacific Northwest and Shanghai, China, is 14 hours. While senior executives overseeing the manufacture of Xbox consoles, Surface computers, and other Microsoft hardware products get their day started, it's the wee hours of the following morning 5,700 miles away. The need to view real-time information and insights is crucial, for both the headquarters in Redmond, Washington, and plant managers in Shanghai.

When Mark Klinkenberg left Intel to join Microsoft in 2008, the software company was preparing to ramp up its hardware business, which meant ramping up its manufacturing know-how. Back then, data from the assembly line sauntered into various databases and spreadsheets. Analysts crunched the numbers and produced reports for a conference call the following day with plant and executive management. The data was old and not always actionable. Mark and his team began to ask, "How can we get smarter faster?" The team needed to look for patterns, relationships, and causality across the plant with greater speed, accuracy, and efficiency in order to develop better solutions for building products and improving yields.

"We had a lot of information," Mark recalls. "But how could we get smarter?"

AI became a critical tool for Microsoft's manufacturing team in China and the United States. Together the team executed on three strategic fronts: get connected, become predictive, and grow to be cognitive. In other words, they needed to connect as many devices, instruments, tools, and people as possible, build algorithms that could convert the massive amounts of data this connectivity enables into predictive insights, and ultimately train the model to be cognizant—to think—of solutions to augment human capabilities.

The goal was to rush data-based insights to people on the plant floor.

They lit up Microsoft's Power BI business analytics to produce data visualizations and dashboards that ensured plant managers and supervisors alike were seeing the same information at the same time—one source of truth. The dashboards funnel all the telemetry data that's collected, such as cycle rate and the pressure of a particular press, into a model to detect shifts and drifts so they can respond before a part fails or a line has to shut down. While some plants do this for a single machine, Microsoft created a system that looks at all machines at once. Finally, the team also created AI for planning purposes. Using a range of statistics and information from the sales team, inventory, promotions, and other data, the team also can have more accurate demand forecasting.

Mark laughs when he retells the story that frontline factory workers started to ask if he could pause data before it was sent to executives. They worried about getting yelled at prematurely, but executives learned quickly that everyone was acting urgently on the same data.

They've been able to move from 40 percent orders committed within five days to 95 percent committed within 48 hours. There's been a $50 million reduction in errors and omissions year over year, and they've saved another $10 million in just one year due to scrap reduction, yield improvements, and process optimizations. Demand forecasts are 15 percent more accurate. But one of the greatest benefits has been a more engaged workforce from top to bottom.

Mark and his colleague Darren Coil warn that the cultural transformation needed for AI to grab hold in manufacturing cannot be underestimated. In fact, they note, the culture will fail before the technology does. The cultural change won't happen through a webinar or reading a manual. It must be a combination of top-down and bottom-up change. In their case, the corporate vice president in charge decided to lead by example—executive role modeling.

No more static data presented in Excel or PowerPoint. All presentations and all meetings took place gathered around live data and analysis using Power BI. Whether you were a factory worker in China or a business manager at headquarters in Redmond, you had to use Power BI.

"Show them, don't tell them," Darren recalls. "Reinforce the lessons and keep coming back. Tell what you're going to tell them, tell them, and tell them what you told them—repetition."

What Darren and Mark have observed is that executives need to set expectations and role model as a first step. It's also critical to involve workers in the process so they can see how the machine learning and AI algorithms are trained. This builds trust and buy-in. Helping them understand these systems will give you better information to make better decisions. Then, the executive needs to step away and come back a few months later to see if the transformation is taking hold. In Darren and Mark's example, the executive returned to find that the factory team was using and implementing data and AI to solve problems and have quicker impact. In fact, they started to become keen consumers of data, demanding access to more data and information—a flywheel or virtuous cycle begins to gain momentum.

A cautionary tale: Mark and Darren point out that one obstacle to cultural transformation is that workers are accustomed to having their performance evaluated using the earlier system. A plant manager might say, "The old system showed me running at 95 percent yield. That was my goal and I was rewarded accordingly. The new system and data show me at 92 percent." Clear communication during the transformation is important. Machine learning and AI data models will be more precise, and senior leaders and workers must be understanding and flexible during this change.

CHAPTER 2

AI AND THE MANUFACTURING WORKFORCE

AI AND THE MANUFACTURING WORKFORCE

On a frightfully cold January day in 2019, just over a year before the U.S. presidential primary, New Hampshire voters were treated to a parade of potential presidential candidates making their case. Headlines about their auditions appeared in Twitter feeds next to President Trump's excitement that "last year was the best year for American manufacturing job growth since 1997, or 21 years … and it's only getting better."

But newspapers were also reporting that manufacturers might be cutting back. The Wall Street Journal reported that technological change sweeping the auto manufacturing industry is forcing job cuts.[9]

The previous summer, a union representing culinary workers in Las Vegas, Nevada, issued a press release noting that it would be negotiating "language regarding increased technology and the effects automation has on jobs and workers." A Huffington Post article reported that "workplace experts say unions need to figure out how to help workplaces and workers adapt to new technologies to reduce layoffs if workers are to have hope of surviving and even thriving in the face of this threat."[10]

At UNI Global Union, an organization representing 20 million workers from more than 150 countries, the issue of ethical AI and the implications for the workforce is something they have been very active and vocal on. In 2016, they called for the creation of a global convention on the "ethical use, development and deployment of artificial intelligence, algorithms and big data."[11]

More recently they issued 10 principles for ethical AI as it relates to the future world of work. According to their report, while some workers are already losing their jobs, they also state that "AI, machine learning, robotics and automated systems can also benefit workers." UNI Global Union called on companies to work with labor and encouraged those companies to follow principles such as demanding that AI systems be transparent, be built with an ethical black box, adopt a human-in-command approach, and ensure genderless and unbiased AI.[12]

Meanwhile, the Organisation for Economic Co-operation and Development (OECD), a group of 36 countries accounting for 80 percent of world trade, recently observed in a set of recommendations that AI may have disparate effects within and between societies, including the deepening of income inequality and widening of the skills gaps.[13]

In this chapter we explore the implications of factors ranging from economic dislocation and workplace safety to the future of work as they relate to manufacturers and policymakers alike. Just as we did in the previous chapter, we begin with the voices of our customers and partners and those on the frontlines.

A NEW RELATIONSHIP WITH TECHNOLOGY: CREATING A SAFER AND MORE SATISFYING WORKPLACE

All the manufacturers we interviewed agreed that worker safety must continue to be the highest priority. After all, much of the momentum behind robots is driven by a desire to replace human laborers doing repetitive and often dangerous tasks.

But even with these advances, manufacturing remains prone to safety challenges. For example, in the United States, muscle sprains, strains, and tears are the leading type of injury in manufacturing, resulting in an average of 10 days away from the workplace for those suffering from these injuries.[14] In Europe, material-handling equipment like forklifts can cause accidents, and accidents cost lives and money—€5 billion per year according to one estimate. As one customer told us: "It can be unsafe today, which is why we are so keen to move to autonomous systems. Our top priority is to get to zero accidents." Another customer says he wants AI that can improve safety monitoring and control in real-time fashion, to make sure equipment is being run as intended. He wants cutting machines that are more sensitive and forklifts that know when people are nearby.

It would be hard to find someone more steeped in the field than Bazmi Husain, a 38-year veteran of Swedish-Swiss technology giant ABB, which specializes in robotics and discrete and industry automation technologies. Listening to Bazmi is rather like attending a TED talk. The future he envisions is one where AI and robotics augment human capabilities to free us from dangerous occupations and repetitive tasks while delivering unprecedented improvements in living standards.

"The only thing I fear about AI is that people will fear AI," he remarks, clearly concerned that trying to put the brakes on AI would be wasting an opportunity to improve people's lives.

ABB pioneered the modern robotics revolution in 1974 with the first microprocessor-controlled industrial robot, and Bazmi has been part of the evolution of smart robotic solutions since he joined the company in 1981. Today, as ABB's Chief Technology Officer, he is focused on innovation, with particular emphasis on AI for industrial applications.

"Robots used to be unaware of their surroundings, so we had to put them in cages or behind fences to keep humans out of harm's way," Bazmi says. "But with advances in sensor technology and industrial AI, robots are now able to sense their environment—to 'know' what is going on around them—which means we have been able to free them from their cages to collaborate with humans. The result has been safer working environments and far higher productivity."

"The only thing I fear about AI is that people will fear AI."

AI AND THE
MANUFACTURING
WORKFORCE

ABB's automated control systems react to planned events and are there to help people be more efficient and productive. But as with all programmable systems, there comes a point where it no longer pays to automate—where conditions become too complex or unpredictable to make programming feasible or worthwhile—and humans have to step in.

"With AI, we can address this challenge," Bazmi explains. "Rather than blindly following rules, AI is able to learn from prior experience and human intervention and so react to unforeseen situations in the most appropriate way. Not only does AI take the pressure off people—reducing the potential for human error—it augments human capabilities by allowing people to focus on tasks that humans are ideally suited to doing, supported by autonomous systems."

ABB is working on autonomous systems for a variety of industries. In December 2018, the company achieved a breakthrough toward autonomous shipping when ABB's new intelligent autopilot enabled a ferry captain to remotely pilot a passenger ferry through a test area in Helsinki harbor.

With AI, ABB can better predict when a machine, robot, or system is at risk of malfunction. For human operators, knowing when to intervene preemptively avoids costly downtime and unpleasant surprises. To appreciate just how unpleasant an unexpected breakdown can be, imagine what happens when an offshore wind turbine suddenly stops working. A standard wind turbine is a 116-foot blade atop a 212-foot tower with a total height of 328 feet—30 or more stories in the air—and wind turbine speeds reach 200 mph. Without AI, operators have no idea when they might have to send a crew out to sea to repair the turbine. In bad weather, it might be days before action can be taken.

With sensors, data, and machine learning algorithms, ABB can not only predict with 95 percent accuracy when a turbine is likely to fail, but the company can provide prescriptive information on remaining useful life under different operating conditions. Armed with this information, an operator can schedule preemptive maintenance at the most opportune moment.

With AI to augment human capabilities, ABB can make its industry safer, cleaner, and more productive than ever before to the benefit everyone.

THE TALENT SUPPLY CHAIN: A SHARED AGENDA FOR SKILLS AND EMPLOYABILITY

The debate around AI-induced automation and the workforce is extremely vivid, but it's rarely based on strong empirical evidence, according to a recent LinkedIn study.

LinkedIn, a Microsoft subsidiary, is a professional social network with nearly 600 million members worldwide. LinkedIn creates what they call "The Economic Graph," a digital representation of the global economy based on data about those members, 30 million companies, 84 thousand schools and universities correlated with 50,000 skills and 20 million open jobs.[15]

LinkedIn's research into emerging skills around the world has shed light on a few growing trends. First, AI skills are among the fastest-growing skills on LinkedIn, and saw a 190% increase from 2015 to 2017. Second, industries with more AI skills present among their workforce are also the fastest-changing industries. Finally, in the same time period, when all the different skills held by an industry's workforce are compared, it's clear which industries have changed the most by how much their overall skills makeup has changed.

The manufacturing sector ranks fifth behind software and IT services, education, hardware and networking and finance for having the most need for AI talent.

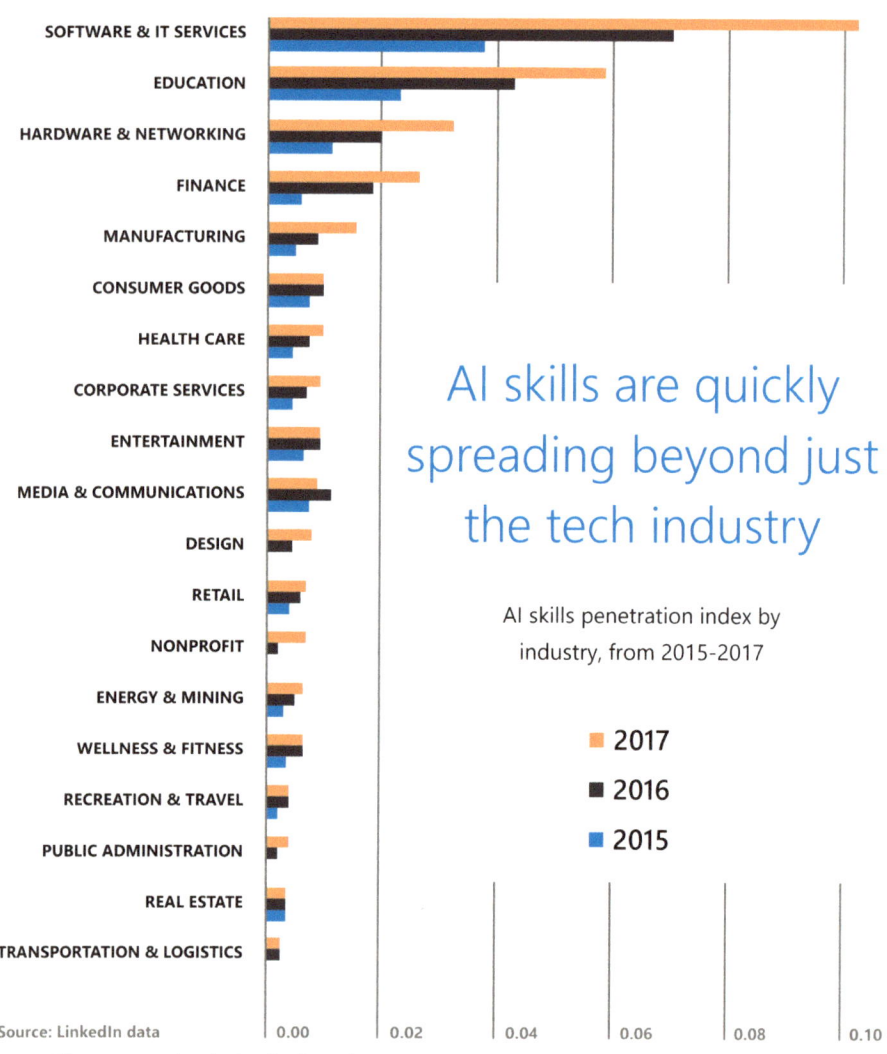

Note: The penetration index looks at how many of the AI skills appear among top 30 skills for each occupation in each industry. It computes the mean % of AI skills across occupations within each industry in year 2015, 2016 and 2017. By comparing the index across years, it gives us the time trend of industry-level AI skills penetration.

This is important data as it demonstrates that despite the huge demand for AI skills in manufacturing, the talent supply is not keeping up.

Researchers at LinkedIn also point out that governments are currently devising their own AI strategy, craving for more objective data and insights about AI progress and hindrance.[16]

In the United States, Jay Timmons, the President and CEO of the National Association of Manufacturers (NAM), has thought a lot about the future of the workforce and the implications for the constituency he represents. His conclusion?

"Our shared belief [is] that every one of us can contribute to the success of our companies, our communities, and our country. Manufacturing is more than just technology and machines; our industry is about people and the potential we can unleash."

According to a NAM survey, 82 percent of manufacturing executives indicate they believe the skills gap will impact their ability to meet customer demand. 78 percent believe the skills gap will impact their ability to implement new technologies and increase productivity.[17]

As a result, NAM has explored a range of policy recommendations, including the need to adapt to the changing needs of the modern manufacturing workforce and the changing attitudes individuals have toward their work. The organization recommends building a system that delivers a sustainable pipeline of strong, mid-skilled manufacturing talent.

The question of how best to prepare workers for the future occupies labor and trade unions everywhere, including Scott Paul, President of the Alliance for American Manufacturing (AAM). AAM is a partnership established in 2007 by some of America's leading manufacturers and the United Steelworkers union.

Scott notes that conversations about workforce reskilling or upskilling tend toward two camps: the "philosophical salons of influential coastal cities" and the more practical factory floor discussions in manufacturing towns. Both are valuable, but increasingly we need to get more practical.

It's a challenge Microsoft's Portia Wu thinks a great deal about.

Before joining Microsoft as Managing Director of U.S. Policy, Portia spent her career developing and implementing labor and workforce policy, including serving as Assistant Secretary for Employment and Training at the U.S. Department of Labor.

"Within manufacturing, we're in a middle chapter, not at the beginning," she explains. "The industry has undergone a lot of change already from robotics and automation."

> **"Within manufacturing, we're in a middle chapter, not at the beginning," she explains. "The industry has undergone a lot of change already from robotics and automation."**

When studying the issue of labor displacement and new skill development caused by technological advances, she notes that policymakers need to clearly identify which workers are being discussed. For example, needs may be very different for workers who are being displaced, for workers whom companies are helping to gain added skills, or "upskill," and for workers who program or design AI systems. There are other important dimensions to consider, including whether workers are just entering the field, or if they are mid-career or late career. Depending on the circumstance, policy solutions may differ. To improve our training pipeline, Portia suggests some key steps are needed:

First, business and industry need to define the needed skill sets that map to the jobs and careers of the future. Where is the demand, and how can we standardize job titles and descriptions? In the past, the profile of a welder was very clear; but, as the pace of technology change continues to accelerate, the role of a data labeler or data analyst may be very different from company to company. Moreover, new jobs and roles will be emerging every day.

Second, these insights need to be shared with policymakers, educational institutions, and, most importantly, workers so that everyone is operating with the same information.

Next, education and training opportunities must be refocused so they can meet those skills needs. This means not only degree programs, but also certificate programs, online learning courses, and internships and apprenticeships. All of these must align with these skills and jobs.

Finally, these efforts need to be evaluated in real time in order to share results and adjust as necessary. The goal should be helping to transform people and their ability to succeed in the jobs of the future as those jobs are created.

Across town from Portia's office in Washington, D.C., is the Siemens Foundation, part of the Global Alliance of Siemens Foundations, whose mission is to ignite and sustain today's STEM workforce and tomorrow's scientists and engineers.

The Siemens Foundation focuses its resources on a particular theory of change: middle skill development for young adults aged 16-30.

Crystal Bridgeman is the Senior Director for Workforce Strategies at the foundation and cites OECD data indicating that Americans rank lower than citizens of other countries on numeracy and problem solving, which limits their job opportunities in STEM occupations like manufacturing, healthcare, IT, and energy. Often what is needed is more than a high school degree but less than a four-year bachelor's degree.

The Siemens Foundation has set three objectives for its funding and advocacy:

First, it wants to be an accelerator of training models that are backed by evidence or show early promise. In particular, this means work-based learning and apprenticeships. The foundation has partnered with the National Governors Association (NGA) in 17 states and with the U.S. Department of Labor through the American Apprenticeship Partnership, among others.

Second, their effort is designed to support training and development initiatives that emphasize high-quality, digital skills and that blend both classroom study and hands-on application in the workplace.

Finally, from an advocacy perspective, the foundation is working to change American perceptions from a "college or nothing" mindset to one that embraces a variety of educational pathways that lead to satisfying careers.

Apprenticeship programs in Europe, especially in Germany and Switzerland, serve as an inspiration for a more robust post-secondary system. There, the idea of an apprenticeship is not a point of contention, Crystal observes. Government, business, and industry are all on board. Apprenticeships are transferrable to higher education, and higher education is transferrable to apprenticeships. To truly emulate the European model will be difficult in the United States, where education policy is set much more at a state and local level than at the national level.

"We need an ecosystem that brings together the right partners and high standards that are measured against valid benchmarks—industry and academic standards," she said.

> **"We need an ecosystem that brings together the right partners and high standards that are measured against valid benchmarks—industry and academic standards," she said.**

Across the Atlantic, the experience with apprenticeship is very encouraging. Unemployment after apprenticeship is very low in Europe compared with unemployment or underemployment following university graduation in the United States.

This is the assessment from Michel Servoz, Senior Adviser on AI and Robotics to European Commission President Juncker. Michel notes that an apprenticeship not only develops skills but is also a great introduction to the world of work, including the necessary relationship-building and communications skills that are needed.

His view is that rethinking education is job number one for policymakers everywhere. He sees a skills mismatch—both technical and soft skills—that if not resolved will lead to grave economic consequences.

Not only do students need technical and digital skills, but they also need to develop the skills that are not likely to be automated anytime soon—creativity, problem-solving in an unpredictable environment, and empathy.

Moreover, education must be lifelong. The current model of education, career, and retirement no longer works. Training and education will become a lifelong pursuit.

Those countries where this is happening are seeing the rewards.

Jobs that once left for low-wage countries are beginning to return. Adidas closed plants in Malaysia and Vietnam and opened plants in Germany and the United States. Philips did the same, returning jobs outsourced to Asia back to The Netherlands.

"There is huge anxiety about the future of work among trade unions," Michel said. "Productivity and wages are stagnant, so they are rightly concerned."

And while these concerns about the impact of technology are universal, Scott Paul, from AAM, says, "Tech is a fact of life. We are not luddites. We're not standing in the way. But there is also a realization that it requires a different set of skills. It's simplistic to say AI will kill manufacturing jobs, but it will transform them."

While there is agreement on its inevitability and benefits, workers are concerned about what companies will do with the savings that result from optimization. Will they reinvest in the workforce? Will they look to the future and make plans to compete in new markets? Or will they simply take the profits? Workers and their unions worry that they will not be part of the conversation as advances are being considered and deployed.

Brad Markell from AFL-CIO Industrial Union Council—who is also a long-time member of the International Union, United Automobile, Aerospace, and Agricultural Implement Workers of America (UAW)—calls this "procedural justice," the notion that you must involve workers in the creation of processes and technologies in order for those processes or technologies be truly helpful to those workers. No one, he notes, wants to end up with "bargained acquiescence," which Black's Law Dictionary defines as "consenting without any enthusiasm."

"You want workers who say, 'I like my job,'" Brad advises. "You want workers who say, 'I am challenged rather than just doing routine things over and over.'"

Brad points to General Motors and Ford as examples of companies that are investing in training and smart factory simulation to help bring workers along. For example, 3D printing is being experimented with for toolmaking and production of small fixtures for autos.

"It's not a peanut butter approach," Brad said. "It's application by application."

Another green shoot—a harbinger of things to come—is a promising lab at the Digital Manufacturing and Design Innovation Institute (DMDII) in Chicago.

> **"You want workers who say, 'I like my job,'" Brad advises. "You want workers who say, 'I am challenged' rather than just doing routine things over and over."**

The lab is a public-private partnership in which workers learn new skills inside a simulated smart factory. While still small scale, the lab demonstrates what's possible when workers learn new digital processes, analytical tools, quality assurance and monitoring, component installation, and how to collaborate with digital and AI engineers to improve systems.

These workers emerge with heightened numeracy and digital skills that are transferable to other factories of the future.

Alongside the huge opportunities that exist to upskill the exiting manufacturing workforce, we may also need to look at new ways to support people for whom disruption means switching to a different career or spending some time out of the workforce. We'll need to adapt new labor laws and policies, many of which were built in the age of the steam engine, not the search engine.

Now a century later, these labor laws and policies are no longer suited to the needs of workers or employers. Health insurance and other benefits, for example, were designed for full-time employees who remain with a single employer for many years. But they aren't as effective for individuals who work for multiple companies simultaneously, such as those who work in the gig economy—think Uber drivers or those who change employers regularly. There is an increasingly pressing need to review these benefits and ensure they provide adequate coverage for all workers in the digital economy, as well as a sustainable contribution structure for business.

WORKING TOGETHER

So, what does all this tell us? Should we be excited or concerned? Perhaps it tells us we need to be a little bit of both.

We should be excited for the opportunity that AI brings to help create more compelling jobs in manufacturing, jobs that are working alongside AI rather than being replaced by AI. That AI will reduce the number of low-value, repetitive, and, in many cases, dangerous tasks. This will provide the opportunities for millions of workers to do more productive and satisfying work.

But we also need to be concerned. Concerned that we have the right training and education infrastructure to create a future workforce that reflects the future workplace and a social safety net that protects all workers in our modern economy.

If every manufacturer will be using AI as part of its operations, then the talent needed will be different in the future than it is today. We need to develop new approaches to training and education that enable people to acquire the skills that employers will need as technology advances and create innovative ways to connect workers with job opportunities.

As we move forward, it will be essential for a new coalition of interests to come together to explore how to best support workers, create economic opportunity, and plan for the future. But this coalition of interests needs to do more than just focus on the workforce, as critical as that is. This coalition needs to help shape the broader ethical and policy frameworks that will foster responsible innovation. The next chapter looks at how we might achieve this.

CLOSING THE AI SKILLS GAP: MICROSOFT'S COMMITMENT

Microsoft has a shared goal with our customers to identify and develop the right set of skills they need to leverage the potential of AI in their business. We are doing this by implementing skills-building solutions across education systems, public agencies, and workforce industries to provide a breadth of skills training to meet the need of individual learners and support countries in building their talent pipelines. So, how is Microsoft helping address the skills gap and enhancing employability for our customers?

First, we started by interviewing and researching organizations who successfully leveraged AI as well as those whose AI initiatives were unsuccessful. Additionally, we spoke with executives familiar with the AI initiatives of companies that are considering the introduction of AI into their organizations. The organizations we explored included companies in the finance, real estate, education, manufacturing, healthcare, and retail sectors, including: Severstal, Nissan, Goldman Sachs, Bloomberg, MI6, Hirotec, Arizona State University Skysong Center for Innovation, University of Kansas Medical Center, Rolls-Royce, Burberry, Walmart, and Tesco.

Our work surfaced several key takeaways. Most fundamental is agreement that there is a lack of AI-related skills in the marketplace and that demand for data and AI skills will continue to outstrip supply. The research also revealed agreement that we are in the early phases of this digital revolution and that the skills and capabilities required today will be table stakes in the future.

Another key takeaway from our research is that there are three core disciplines that must be present and active in every AI project to be successful: Data Science and AI, Data Management, and AI Business Integration. We considered including a fourth discipline,

Infrastructure Management, as none of the other disciplines will succeed without a robust, technical environment supporting Data Management and Data Science and AI. Each of the disciplines has its own track of learning curricula, and this is how we have designed our online learning programs that span the entire journey from core foundational knowledge to specific technical skills needed in the workforce.

Armed with this knowledge from our customers and partners, we created a suite of programs that are available worldwide to address the skills needs:

The Microsoft Professional Program (MPP): The MPP was created to help people gain technical job-ready skills and get real-world experience through online courses, hands-on labs, and expert instruction, which culminates with a capstone project that allows learners to demonstrate their learning. Throughout the program, students receive certificates that accrue to a Certificate of Completion for the entire MPP track upon completion of the capstone. Each track is aligned with a tech job role in today's workforce. Because MPP learners are required to demonstrate proficiency in the skills they acquired, employers have the confidence they need to hire skilled individuals for in-demand roles, which puts potential candidates onto the path to a rewarding career. Among an extensive list of tracks, the MPP offer tracks in Data Science, Artificial Intelligence, AI Developer, and Data Analyst.

Learning as a Service (LaaS): LAAS helps our customers with Azure-based learning solutions to train their employees or students on the latest Microsoft technologies. This program combined with MPP provides customers with differentiated and tailorable, technology-neutral skill building in high-demand job areas such as AI Engineering, AI App Developer, and more.

The Microsoft Imagine Academy (MSIA): Reaching over 8 million students and educators annually as of 2019, MSIA provides the industry-aligned curricula and certifications to build competencies and validate skills for high-demand technologies. MSIA offers courses and certifications in four paths: Computer Science, IT Infrastructure, Data Science, and Productivity.

LinkedIn Learning: This platform combines an unmatched library of more than 13,000 courses taught by real-world experts with LinkedIn data and insights drawn from over 575 million member profiles and billions of interactions, which gives employers a unique and real-time view of how jobs, industries, organizations, and skills are evolving, while also helping leaders identify the skills their organization needs to succeed.

AI Business School: This online course is a master class series that aims to empower business leaders to lead their organizations on a journey of AI transformation. Designed specifically from a nontechnical perspective and created in collaboration with INSEAD (a graduate-level business school with campuses in Europe, Asia, and the Middle East), the course materials include brief written case studies and guides, plus videos of lectures, perspectives, and talks that executives can access online. A series of short introductory videos provide an overview of the AI technologies driving change across industries, but the bulk of the content focuses on managing the impact of AI on company strategy, culture, and responsibility.

Microsoft Philanthropies: By 2020, more than 800 million people will need to learn new skills for their jobs and two-thirds of students today will work in jobs that do not yet exist. Not only does this skills gap impact prospects for individuals, it has a systemic effect on the ability of companies, industries and communities to realize the full potential of this digital transformation. Addressing this issue involves changing the way people are educated and trained, and the way companies hire and support their employees. Reshaping the labor market for the 21st century is an enormous task that is bigger than any one company. That's why Microsoft Philanthropies is working with businesses, governments, nonprofits, and educational institutions to help those already in the workforce, those trying to re-enter the workforce or build new skills, those who are seeking to gain work experience, as well as equipping future students with the digital skills critical to ensure both financial stability and opportunities for growth.

Microsoft is leveraging its position as a leading global technology company to develop technological solutions, and make strategic investments in skills development and employability programs, as we work to ensure that this industrial revolution and the technology driving it, creates economic opportunity for all. Microsoft Philanthropies partners with nonprofits to increase equitable access to high quality computer science education, specifically seeking to reach young people who are least likely to have access to it. Around the world, more than 80% of the young people who benefit from Microsoft Digital Skills grants come from underserved communities, and more than half are female. These efforts support our commitment to ensure all people have the opportunity to learn the skills they need to succeed in an AI-enabled world.

CHAPTER 3

FOSTERING RESPONSIBLE INNOVATION

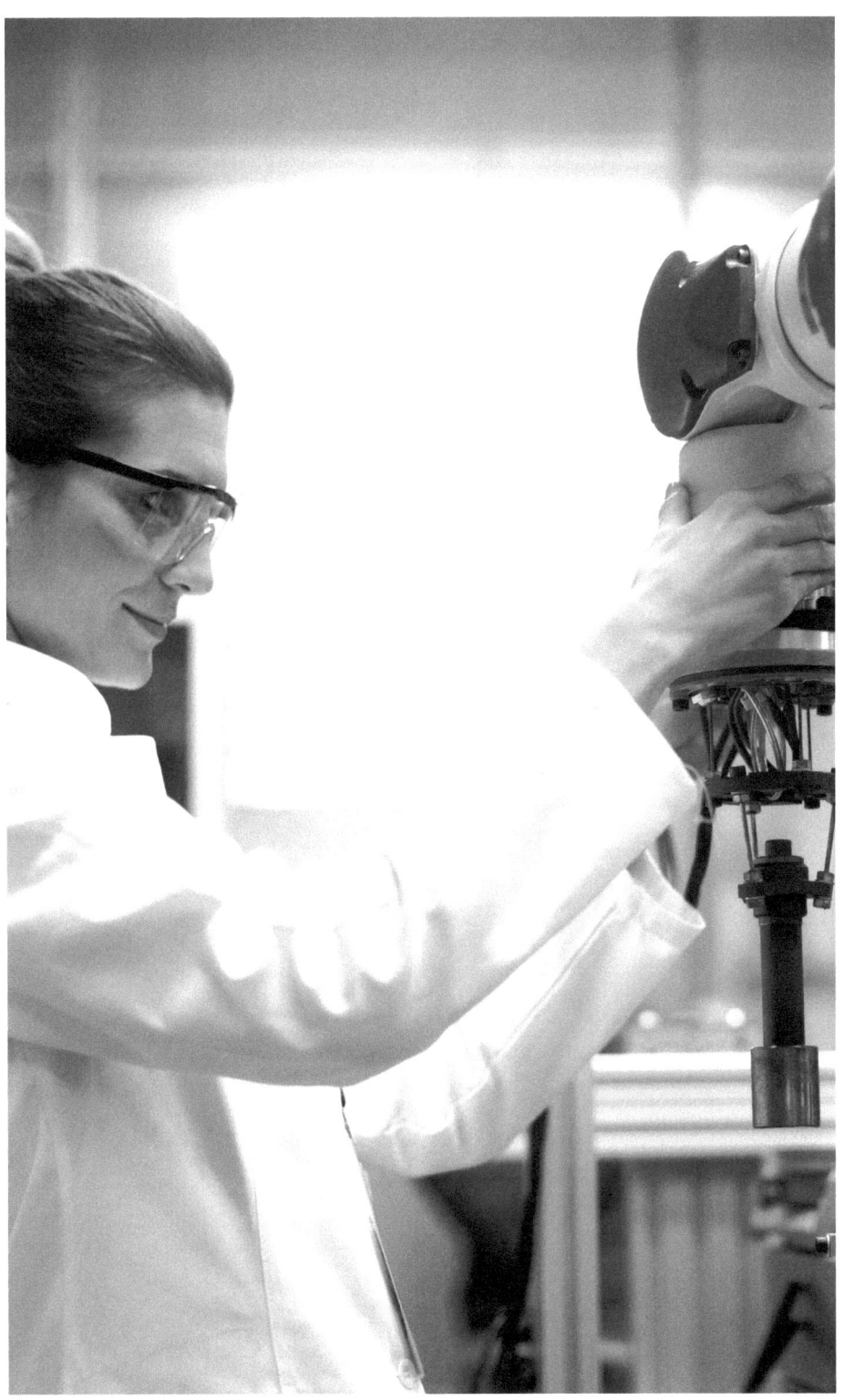

FOSTERING RESPONSIBLE INNOVATON

Most days on the Microsoft campus outside Seattle, policymakers mingle with manufacturers and representatives from other industries in the company's Executive Briefing Center, where demonstrations of AI, drones, robots, augmented reality, and other advanced technologies are on display. What's on display is not futuristic: It's here today.

One example is a Minnesota family's corn farm. Neatly laid out on a 3D topographical map, the demonstration of the farm shows how weather, seed data, satellite imagery, historical growth, soil moisture, and other variables inform farmers how to optimize corn production for companies like Land O'Lakes and Purina.

One of the Microsoft demonstration teams points out that some of the most technologically advanced companies look on in disbelief, smiling and asking themselves aloud, "How are we being beat on AI by a bunch of farmers?" But those in AI-informed farming respond that, at the end of the day, it's the farmer—not the machine—that is entrusted with the ultimate decisions.

In these examples, AI solutions help the farmer, first and foremost. It's the farmer, using machine learning to decide against a fungicide in preference for a John Deere planter that can put down variable seeds based on which strain of seed works best in what soil. AI is the tool of the farmer.

Next to the farming demonstration is the actual cockpit of a Singapore Airlines passenger jet using a Trent 1000 Rolls-Royce engine for the Boeing 787. Singapore Airlines came to Microsoft asking for AI and machine learning to accurately prescribe the fuel levels for each flight—the optimal level that takes into consideration weight, wind, distance, and engine efficiency. Rolls-Royce is extending its manufacturing business into engines-as-a-service for prescriptive maintenance and operation.

More than 25 sensors track engine health data, air traffic control, route restrictions, and fuel usage to uncover data insights that will enable airlines to improve operational performance and fuel efficiency.

In demonstration after demonstration, the conclusions are obvious: A thriving manufacturing sector and the use of the latest technologies are inextricably linked, and that a thriving manufacturing sector is a priority for the broader economy.

In fact, according to a report by the World Economic Forum (WEF) on manufacturing and the future, manufacturing has the largest multiplier effect of any economic sector.[18] The WEF report points out three important areas for policymakers, business, and civil society alike:

1. **Public- and private-sector interests need to converge:** This highlights the need for new industrial policy collaboration models, starting with new skills development models.

2. **A new manufacturing language must focus on capabilities and global value chains:** To better anticipate underlying constraints on industrialization, we need to move beyond the typical gross domestic product measures.

3. **New, connected business industrial models:** This can help prepare for the potential Fourth Industrial Revolution.

While the WEF report also suggests these areas require further discussion and that further research between the public and private sectors is needed, it does pose some immediate questions: How can further convergence of the public and private sectors, as well as civil society, be promoted? How can multiple policy initiatives and instruments be aligned in one direction (e.g., supplier development programs, promotion of foreign direct investment, upgrading infrastructure, skills development)? At what pace have countries and businesses adapted to the changing manufacturing environment? How should industrial policies be focused to match the development of future capabilities? How can countries effectively move up the value chain?

These questions are not easily answered, particularly given the pace of change. Moreover, it is often challenging to make definitive decisions about policy and technology when you are in the midst of an industrial revolution.

The countries that thrive in this climate are those that enable the right framework of laws, policies, and principles to realize the potential of these new technologies while also curbing their negative effects. In the era of the Fourth Industrial Revolution, AI is no different.

But, as we have witnessed with previous technology paradigm shifts—from the plow to the harvester, from the carriage to the automobile, from the mainframe to cloud computing—nations have prospered when they address these issues earlier, even if the solutions aren't immediately clear. The countries that thrive in this climate are those that enable the right framework of laws, policies, and principles to realize the potential of these new technologies while also curbing their negative effects. In the era of the Fourth Industrial Revolution, AI is no different.

For our part, we believe that AI will be the defining technology of our time. Like the discovery of electricity or the development of the steam engine, AI will have the power to fundamentally change people's lives, transforming industry and transforming society. It is therefore even more imperative that the development and deployment of this technology adopt clear principles to govern its creation and use.

In his book, *Hit Refresh*, Microsoft CEO Satya Nadella writes that just as our ethics, values, and laws have been developed and evolved over generations for the physical world, so too must our understanding and rules for the digital world. He writes, "The most critical next step in our pursuit of AI is to agree on an ethical and empathic framework for its design."[19]

Through the experience of our customers, we've explored what AI is accomplishing in the manufacturing supply chain today and the implication for the talent pipeline, and we also need to think about AI's broader societal and ethical implications. This chapter explores these implications and public policy considerations that can help create the right conditions for responsible innovation.

ETHICAL AI IN MANUFACTURING

Microsoft has listened carefully to what a diverse group of stakeholders in manufacturing have to say. From workers on the factory floor to executives in the C-suite, and from labor unions and academics to industry advocacy groups, what has come through all these conversations is a sense of shared responsibility to create trusted, responsible, and inclusive AI systems. We've heard that it's time to work together to reach a consensus about what principles and values should govern the development and use of AI in manufacturing and that these values need to reflect the experiences of organizations that are using AI every day.

Back in early 2018, in *The Future Computed: Artificial Intelligence and its Role in Society*, we wrote that designing trustworthy AI requires creating solutions based on ethical principles "deeply rooted in important and timeless values."

Source: Microsoft Corporation, *The Future Computed: Artificial Intelligence and its role in society*, January 2018

In the book we proposed some principles we thought would assist in guiding the ethical development and deployment of AI. These six principles called for AI systems to be, reliable and safe, private and secure, fair, inclusive, transparent, and accountable.

At the time we published these principles, the discussion on AI ethics was still in its relative infancy and there was not what we would characterize as a deep discourse on what responsible AI might mean for specific sectors such as manufacturing. Back then, the most debated topic focused on autonomous vehicles: How safe and reliable are the systems? How do they make decisions and on what basis? What data do they collect about the driver? Who's liable when things go wrong?

Today, however, the discussion has somewhat matured. The debate on ethics and AI is no longer limited to the end product of manufacturing like self-driving cars—it now covers the entire digital value chain from design and engineering, planning, supply chain management, factory automation, and workforce training to IoT.

During our exploration of AI in manufacturing, we asked leaders about their attitudes toward the concepts of responsible innovation and ethical AI. Very often, unprompted, manufacturers said, "We need to be extremely ethical." Every one of them was deeply thoughtful about this issue.

Their responses raised a spectrum of issues from the risk of premature or overzealous regulation to workplace safety, workforce development, data privacy and security, consumer protection, and product liability laws.

Equally, much of the conversation revolved around the machine-to-human relationship and ethical considerations about keeping "humans in the loop."

We were surprised and encouraged by the depth of our customers' thinking about AI's ethical dimensions, and their responses have prompted us to think carefully about how the six principles for ethical AI might apply to manufacturing.

Indeed, when we developed these principles our thinking was very much focused on our own role as a creator of AI technologies and how our AI products and services could win the trust of customers and policymakers. But as we conducted the research described in this book, it solidified our assumption that these same principles should and do apply to any organization that uses AI, and, in particular, how these principles might apply to the manufacturing sector.

And while we believe the principles are broadly applicable, we also recognize they will have different weight and relevance in different segments of the manufacturing sector. For instance, the principle of privacy will have more relevance for the producers of smart elevators than the producers of milk cartons.

Moreover, these principles are not about meeting some rigid checklist; they are about providing a flexible framework to promote the coexistence of business value and ethical values. As AI systems become more mainstream, we begin to understand that society has a shared responsibility to create trusted AI systems and that collaboration is essential to reaching consensus about the principles and values that should govern AI's development and use.

Let's explore how each of these six principles might be applied to the manufacturing sector.

Safety and Reliability

Perhaps the most critical principle in manufacturing scenarios is safety and reliability. Every customer we spoke to put employee safety and product reliability at the very top of their list—the complexity of AI has fueled fears that it may cause harm in unforeseen circumstances. We see this concern arising in areas ranging from autonomous robots on the factory floor to autonomous vehicles on public roads.

As with any technology, however, trust will depend on whether AI-based systems can be operated reliably, safely, and consistently over time and not just under normal circumstances—they need to be safe, predictable, and reliable in unexpected conditions or when they are under attack. In manufacturing, concerns about safety and reliability take on a more urgent tone, especially as factories become increasingly digital.

In a world where factory equipment is increasingly more complex, automated, and interdependent, it's not surprising that many of our customers see the opportunity to harness AI to make their workplaces safer and their processes more reliable. AI is increasingly important in predictive maintenance for equipment, with sensors tracking operating conditions and performance of factory tooling, predicting breakdowns and malfunctions, and taking or recommending preemptive actions.

So, when developing AI systems for use in the manufacturing process, particular care should be taken to ensure the safety of the people who will be interacting with the system. In addition, AI systems should be trained to function reliably in real-world situations and respond to unforeseen circumstances. For example, an object recognition system that inspects physical parts on an assembly line for manufacturing defects should be trained to function in low-light environments, even if normally it will operate with regular lighting. As another example, autonomous robots that move around the factory floor could be designed to recognize spills on the floor or respond to voice commands in the event of an emergency.

Cybersecurity is also a key component of ensuring the reliability and safety of AI systems. When AI systems are used in the physical world, cybersecurity is critical to protecting people's safety. For example, a cyberattack that incrementally slows the speed of a turbine on a cooling system could result in overheating of the plant and a major safety incident. Similarly, a facial recognition solution for monitoring employee access to sensitive areas of a manufacturing facility needs to have robust measures in place to protect against malicious interference.

Beyond the factory floor, cybersecurity is critical for the safety of consumer products that have AI embedded. The stories of hacked baby-cams are certainly concerning,[20] but cyberattacks that compromise the software in digital pacemakers or automatic insulin pumps can mean the difference between life and death. Regulators, such as the U.S. Food and Drug Administration, are becoming more aware of this issue and are strengthening the cybersecurity requirements for a range of consumer products, including medical devices.[21]

These issues of safety and reliability bring us back to the ongoing debate about the relationship between humans and machines. AI systems should be designed to be human-centric; namely, so people are always in the loop and can take control when necessary. Indeed, some of our customers not only seek to put humans in the loop but have developed language and terminology about autonomous systems that underscores human primacy over machines.

Ultimately, creating safe and reliable AI is a shared responsibility. It is critically important for industry participants to share best practices for design and development, such as effective testing, the proper structure of trials, and reporting.

There is perhaps no better place to conduct such testing and trials than in the factory setting. The expertise and discipline that manufacturing brings to ensuring a safe workplace and reliable equipment could be essential to learning how AI will apply in the real world.

The importance of safety and reliability goes as much for the factory floor as it does for the end products that consumers and citizens use and interact with. Industry groups and standards organizations can play an important role in helping share and promote best practices and information sharing across the sector. It's clear from our own conversations with customers that they are eager to share their insights and learnings to instill more confidence in the reliability and safety of AI systems.

Privacy and Security

Every stage of the manufacturing process involves humans in one way or another. After all, at the end of nearly every manufactured product is a human user or consumer. But enabling the benefits of AI in applications that affect people requires access to large amounts of data about people.

And just as humans are part of every stage of manufacturing, data is a needed at every stage of an AI solution: from developing and training the AI algorithms to implementing and monitoring the AI system. Some examples from our customers and our own experience underscore the ever-expanding role of data in intelligent manufacturing. But how to protect privacy and secure the personal data that is used in AI systems is a critical issue that is nowhere more important than in manufacturing.

AI systems used during the manufacturing process can directly or indirectly capture personal data about people on the factory floor. AI solutions designed to measure the efficiency and proper functioning of equipment, for instance, may yield information about individual employees' efficiency and performance, while AI solutions that include a microphone functionality (for example to monitor ambient noise or to respond to voice commands) may inadvertently capture employees' private conversations. Therefore, when using such systems, employers should carefully consider the potential privacy impacts on employees and make sure that the use of these systems fully comply with all applicable laws, including employment laws. Engaging with stakeholders, such as employee representatives, can also be helpful to socialize new systems and encourage uptake.

We anticipate that manufacturers will be increasingly using AI and vast amounts of data to drive insights. These insights are not limited to running their manufacturing processes and their supply chains; they will inform them about their end products, the services bundled with these products, and how human users interact with them. Accordingly, when making use of personal data, AI systems must comply with applicable privacy laws, including both sector-specific laws and regional laws such as the EU's General Data Protection Regulation (GDPR), which has broad extraterritorial reach.

Privacy laws generally require transparency about the collection, use, and storage of data, and mandate that consumers have appropriate controls so they can choose how their data is used. AI systems should also be designed so that personal data is used in accordance with established privacy practices and is protected from bad actors who might seek to steal this personal data or inflict harm. In order to facilitate compliance with privacy laws when sharing personal data to develop or deploy AI systems, industry processes should be developed and implemented for the following: tracking relevant metadata about personal data (such as when it was collected and the terms governing its collection); accessing and using shared personal data; and auditing access and use.

Fairness

FOSTERING
RESPONSIBLE
INNOVATION

When we talk about AI and fairness, we mean that AI systems should treat everyone in a fair and balanced manner and not affect similarly situated groups of people in different ways. Understanding why AI systems can behave unfairly and in what ways, as well as understanding who is most likely to be at risk of experiencing unfairness, is fundamental to operationalizing this principle. While there is no single definition of fairness that applies to all AI systems in all contexts, a variety of processes and tools are emerging to help detect and mitigate unfairness throughout the AI development and deployment lifecycle.

Fairness is particularly important where AI systems are used to make decisions that impact people, something that could happen during the manufacturing process in a variety of ways. For example, a company could use an AI system to monitor workers' alertness levels when operating heavy machinery by reading their facial expressions, or to optimize efficiency by providing real-time feedback and insights based on factory workers' location and activities on the factory floor.

In each case the AI solution needs to have appropriate training data, model definitions, and fairness criteria to help ensure that it makes appropriate, fair decisions and operates within appropriate parameters. Training data that doesn't reflect the real world—or systems that inadvertently incorporate biases present in the real world because they are embedded in the training data or inherent in the decisions of the system's designers—are common causes of AI bias. In the example above, a facial expression reading system trained on a dataset composed primarily of images of men, or using model assumptions optimized for male features, might have trouble properly interpreting the expressions of female employees.

AI systems should also be designed to operate within appropriate parameters, so they don't inadvertently make unfair decisions about employees or put employees' safety at risk. For example, an AI system that helps inform performance reviews or employee remuneration should be designed and reviewed regularly to ensure that it doesn't inadvertently take into account factors outside the employees' control (such as gender), doesn't directly or indirectly produce outputs that are based on or influenced by irrelevant characteristics (such as age), and has enough flexibility and "breathing space" built in to allow for workers with different capabilities (such as poor vision but strong dexterity) to be considered fairly. In particular, supervisors or senior managers making decisions based on AI systems should understand the limitations of the system and should not assume that these systems are more accurate and precise—or more sensitive to context—than they actually are.

Fairness becomes even more important when you consider the implications for manufacturing products that have embedded AI solutions.

Fairness becomes even more important when you consider the implications for manufacturing products that have embedded AI solutions. Outside the production phase, fairness is an important foundation of the product design and engineering stage because it is important to consider the ethical implications of products that can make decisions that might affect people in real-life scenarios. For example, the manufacturer of smart doorbell cameras that recognize frequent visitors and flag suspicious activity would want to take particular care to ensure that the doorbell does not make unfair or discriminatory inferences such as those based on gender or race.

Decisions made by teams at every stage of the AI development and deployment lifecycle can lead to unfairness. Therefore, much like security and privacy, fairness cannot be treated as an afterthought or a "bolt-on." Teams should implement processes to identify possible causes of unfair behavior—including societal biases—potential impacts on the people who will use or be affected by the system, and appropriate approaches to mitigate these impacts.

Moreover, to help avoid bias and advance fairness, those designing AI systems should reflect the diversity of the workplace and the marketplace, and these teams should solicit input from diverse stakeholders during the design process. In addition, if the recommendations or predictions of AI systems might be used to inform consequential decisions about people, it is critical that humans are "in the loop" on those decisions and that a designated person or people are accountable for these decisions. Those with accountability must be trained to understand the limitations of the AI system, and to know how to intervene when necessary. In addition, if AI systems are designed to evolve over time, it will be important to ensure that the fairness of the system is kept under continuous review, taking into account the context in which the system operates.

Inclusiveness

Inclusive AI means that everyone should benefit from intelligent technology, especially those with different physical capabilities. The manufacturing workforce today is more socially and racially diverse than many other sectors, but not so when it comes to age, gender, or physical capabilities. For manufacturers, improving inclusiveness means providing the workplace with a range of AI solutions that will allow for more inclusive hiring and retention. This is important in many countries where the demographics of an aging workforce means that those working in the factories or on the shop floor are likely to be more reliant on assistive technologies (e.g., technologies helping aging workers with hearing and vision impairments or mobility limitations). In the United States, the median average age of workers in the metalworking and machine manufacturing industries is close to 49 years, with nearly 60% of the entire workforce over the age of 45.[22] The situation is even more acute in Japan—a manufacturing powerhouse—where in the last 20 years the number of working-age adults has shrunk by 13% and in the next 20 years more than 30 percent of the population will be over 65 years of age.[23] What this means is that if countries like the United State and Japan are to maintain their competitiveness in manufacturing, companies must accommodate an aging workforce with assistive AI technologies.

Thus, when we think of AI in manufacturing, we should not think exclusively in terms of such efficiency-enhancing approaches as factory automation, preventive maintenance, or intelligent supply chains. We should also think about how AI can improve the overall well-being of this diverse workforce.

Two areas to consider are healthcare and ongoing education. Because of its diversity and the nature of the tasks it performs, the manufacturing workforce has healthcare needs that are not the same as those of, say, software developers, business executives, or lawyers. Embedded AI in employer-provided health insurance and wellness programs can provide unique new benefits tailored to the needs of factory employees.

Similarly, as discussed elsewhere in this book, ongoing reskilling and upskilling is now looming as a critical issue for the manufacturing workforce. While in one sense AI can be said to put some manufacturing jobs at risk, it can also be part of the solution by providing new teaching tools to enhance workers' skills and help them transition to new roles.

Manufacturers should consider how their products, including those embedded with AI systems, could be used by a diverse range of people, of different ages, genders, or physical capabilities.

Inclusiveness is also important beyond the factory floor, in the wider society where the manufactured products are used. Manufacturers should consider how their products, including those embedded with AI systems, could be used by a diverse range of people, of different ages, genders, or physical capabilities. Making sure that technology is inclusive and accessible to everyone is important to ensure fair access. In Microsoft's 2018 book, "The Ability Hacks", Microsoft developer Guy Barker implores fellow developers to:

"Consider how each feature might help someone interact with your app. And as you build up your processes to design, build and test products that every person can efficiently use, always keep in mind the rationale for doing so. It's a question of fundamental fairness. No person should be prevented from being connected with others, or from being employed, simply because apps block them from those core aspects of life. And what's more, with the tools at your disposal, it's practical for you as an individual who cares about quality for all to build an accessible app."[24]

Transparency

Underlying the preceding values are two foundational principles that are essential for ensuring the effectiveness of the rest: transparency and accountability. Let's begin with transparency.

When AI systems are used to help inform decisions that have tremendous impact on people's lives, it is critical that people understand how those decisions were made. Transparency improves the quality of products and engenders trust among manufacturers, customers, partners, and regulators, ultimately leading to wider adoption. In part, transparency means that those who build and use AI systems should be forthcoming about when, why, and how they choose to build and deploy their systems, as well as their systems' limitations. Transparency also means that people should be able to understand and monitor the technical behavior of AI systems—what we refer to as "intelligibility."

The need for intelligibility can arise in an almost limitless range of scenarios involving any number of human actors. And the methods for achieving intelligibility can be just as varied. In some cases, information about the development pipeline, the relationships between different system components, or performance metrics may be enough to help stakeholders achieve their goals; in other cases, such as those involving system developers, more detailed information about system components may be required. For example, if an AI system corrects a machine that may be shifting and drifting, manufacturers should be able to look behind the data and algorithm to determine in detail why and how the correction was made. Similarly, for the manufacturer deploying the system, it's not enough simply to receive a notification about a predicted failure—the manufacturer needs to understand how the AI reached that conclusion. Ultimately, it is the manufacturer, not the machine, that is accountable, and for the manufacturer to be accountable, they need to understand how the AI system works.

In other cases, stakeholders such as workers and consumers may need more basic information about the operation of an AI system in order to interact with it in an accessible and useful way.

This builds trust in AI systems and enables people to identify potential performance issues, safety and privacy concerns, biases, exclusionary practices, or unintended outcomes. For example, an agricultural AI solution may be used to help farmers select strains of seeds to plant on the basis of predicted weather patterns and soil quality. The farmers who make use of the AI solution do not necessarily need to understand the technical models underlying the algorithm, but they should understand the factors that the AI system took into account and the limitations of the methodology adopted so they can make informed decisions. For instance, the AI solution may not have accounted for other factors that may be relevant to the farmer, such as whether the seeds were sustainably sourced, the impact of the seeds on the quality of next year's soil, or the health of local pollinator populations.

In short, the people who design and deploy AI systems must be accountable for how their systems operate.

A number of promising approaches to achieving intelligibility have begun to emerge. Some facilitate understanding of key characteristics of the datasets used to train and test models. Other approaches focus on explaining why individual outputs were produced or why predictions were made. Others offer simplified, but human-understandable, explanations for the overall behavior of a trained model or an entire AI system. We encourage further research into these and other approaches to help expand the range of tools available to address the diverse requirements of all stakeholders.

We also believe that those who use AI systems should be transparent about when, why, and how they choose to deploy these systems. This is particularly important when AI systems are used to make consequential decisions, such as selecting suppliers, carrying out performance reviews, or operating key functionalities in consumer products. It is important that people know that AI systems are being used and have information about the purposes and limitations of these systems. Among other things, transparency ensures that ongoing feedback can be obtained from people who interact with AI systems, so that people can get in touch if they see errors or have concerns. This is important to ensure AI systems continue to function properly and safely as well as foster trust in the AI system by the people who use it.

Accountability

The final, and perhaps most important, principle is accountability. In short, the people who design and deploy AI systems must be accountable for how their systems operate.

Manufacturers should draw upon industry standards to develop accountability norms for their own organizations. These norms can help ensure that AI systems are not the final authority on any decision that impacts people's lives, and that humans maintain meaningful control over otherwise highly autonomous AI systems, especially when AI systems make consequential decisions. In order to ensure that people remain ultimately accountable for AI systems and their operation, those who use AI systems on a day-to-day basis should be trained to review results, identify errors, and understand the limitations of the AI system. They also should have the authority and training to take any remediation steps necessary.

Organizations should also consider establishing a dedicated internal review body. This body can provide oversight and guidance to enterprise leaders on which practices should be adopted to address the concerns discussed above and on particularly important questions regarding the development and deployment of AI systems. They can also help with defining best practices for documenting and testing AI systems during development, as well as providing guidance when an AI system

will be used in sensitive cases, such as decisions that may deny people consequential services like healthcare or employment, create risk of physical or emotional harm, or infringe on human rights. At Microsoft, we take accountability very seriously. Our Senior Leadership Team established the AI and Ethics in Engineering and Research (AETHER) Committee to serve in an advisory role on rising questions, challenges, and opportunities with the development anddeployment of AI and related technologies. In addition to its work in support of the principles above, the committee is charged with providing insights and considerations on a broad range of policies, processes, and best practices.

Ultimately, these principles must be reinforced by public policy, laws, and standards to ensure that they are evenly applied and to build both confidence in the technology and trust in those who create it.

NEW RULES FOR NEW TECHNOLOGY

While principles and best practices can be important safeguards in the absence of laws or regulation, it is clear that more formal policy development is also needed to help create the right environment for responsible innovation in manufacturing to flourish.

During our exploration of AI in manufacturing, we asked leaders about their own attitudes on policy and regulatory priorities. Every customer we spoke to was already thinking about the role of policy in shaping the use of AI in manufacturing.

Not surprisingly, their attitudes ran the spectrum of issues that reflects the diversity of the manufacturing organizations they represented. What emerged, however, were a few consistent themes:

- The importance of a regulatory regime that fosters confidence in AI, both by those who invest in AI technologies and those who interact with it.

- Ensure that there is equitable access to AI and, most importantly, access to the critical data that will fuel these AI-led innovations.

- Capacity-building for organizations to leverage AI and the workers who will be working with AI.

- Collaborative approaches to the creation of the rules, standards, and laws that balance innovation with responsibility.

- The importance of a regulatory regime that fosters confidence in AI, both by those who invest in AI technologies and those who interact with it.

- Encouraging the use of AI to create a more environmentally sustainable supply chain.

Below we will explore each of these in more detail:

Fostering Confidence

The direction of policy developments on issues of data privacy and security, intellectual property rights, consumer protection, and liability are universal concerns raised by customers with whom we met.
They wanted a consistent global regulatory framework that fosters confidence in AI for those who invest in AI technologies and those who interact with it. In the area of privacy and security, customers want a consistent regulatory framework on how to share data with confidence, both internally and across the world, particularly given the increasingly interconnected global supply chain.

Customers want a consistent regulatory framework on how to share data with confidence, both internally and across the world, particularly given the increasingly interconnected global supply chain.

FOSTERING RESPONSIBLE INNOVATION

Organizations that invest heavily in AI technologies also want to ensure that they can retain effective control and ownership over the proprietary data that gives their AI systems a competitive edge in the market. As one customer noted: "We [have] legal limitations when it comes to external use of data … [but] there is very little to no limitations inside the plant … it is more of an ethical question in terms of how much we can use to train machines. There needs to be clear regulation that ensures data will be kept private. Ownership of the data is very important. A customer's data stays their data."

In addition, manufacturers also are looking carefully at the future of consumer protection and product liability laws. In a world of data and AI-driven decision-making, who is liable when things go wrong: the operator or the creator of the AI? How can manufacturers, developers, and operators work together to apportion liability to ensure that consumers have adequate protection when something goes wrong, while also ensuring that incentives to innovate are reasonably preserved?

Although frameworks that assess the reasonableness of the conduct of relevant actors are well suited, some frameworks, such as strict manufacturer's liability, may not always lead to the most equitable outcome when AI systems are involved. Since multiple parties can be involved in the development and deployment of AI systems, it will be important to ensure that liability systems reasonably apportion liability for any defects or other harms. It will also be important for industry actors to work closely with insurers to make sure that the risks posed by AI systems are properly insured.

Finally, it should be remembered that AI systems do not operate in a regulatory vacuum. AI systems are subject to a number of existing laws, including privacy laws, product liability laws, consumer laws, anti-discrimination legislation, and sector-specific laws. Existing data-ownership models, such as trade secret protection, may also apply to some aspects of AI systems. Governments should assess existing regulatory frameworks that apply to AI systems and consider both where new measures are needed and where existing legal requirements, or a lack of clarity in what the law requires, pose unnecessary impediments to AI innovation.

Importantly, any new regulatory frameworks on AI should be consistent with existing laws. Industry should work closely with governments to provide practical guidance on the direction for future policies on AI regulation, and developing industry standards can be a good starting point for establishing consensus in the market.

Equitable Access

Democratizing access to AI is a key priority for stakeholders. They remarked that public policy is necessary to create the right incentives to help all organizations, regardless of size, benefit from AI's potential.

They agreed that equitable access to AI, across the entire supply chain, would have a multiplier effect with greater efficiencies, deeper insights, and speed to innovation being shared by all.

Consistent with this concept of equitable access was the key role of data; namely, that if the necessary training data is only available to those organizations that control the largest data estates, then this could disadvantage smaller manufacturers.

As one Microsoft executive explained, "Data are as fundamental to the AI revolution as fossil fuels were to the industrial revolution. Unlike fossil fuels, however, the challenge with data is not scarcity but access."

FOSTERING
RESPONSIBLE
INNOVATION

Perhaps more striking are the comments by experts who worry that the concentration of data in the hands of a few large companies could threaten AI innovation and pose a challenge to equal opportunity more broadly. It is vital that we get the fundamental tools and resources of data and AI into the hands of as many people as possible.

One way to address this information asymmetry is to have policies that expand "data commons" efforts across the manufacturing supply chain. This would provide organizations large and small with access to rich and diverse raw datasets that can be used to help refine and train their AI solutions. Such efforts bring together collaborators to address issues that none of the entities alone can tackle.

Governments can also help accelerate AI advances by promoting common approaches to making public-sector data broadly available for machine learning. A large amount of data resides in public datasets managed by governments that could be exceedingly useful to manufacturers, such as climate data, land survey data, or datasets created from publicly funded research. Governments can also invest in and promote methods and processes for linking and combining related datasets from public and private organizations (on a voluntary basis) while preserving confidentiality, privacy, security, and ownership rights as circumstances require.

Any efforts to create such data commons would need the right data protection and privacy rules, licensing and data sharing agreements, protection for intellectual property and standards, and finally, standardized format or schemas to ensure broad usability.

A recent example of sharing data across the technology industry is the Open Data Initiative (ODI) between Microsoft, SAP, and Adobe.[25] This partnership is designed to improve customer experience management by empowering companies to derive more value from their data and deliver world-class customer experiences in real time. The ODI is a common approach and set of resources for customers based on three guiding principles:

- Every organization owns and maintains complete, direct control of all their data.

- Customers can enable AI-driven business processes to derive insights and intelligence from unified behavioral and operational data.

- A broad partner ecosystem should be able to easily leverage an open and extensible data model to extend the solution.

Based on these principles, every organization owns and maintains complete, direct control of all their data. The core focus of the ODI is to eliminate data silos and enable a single view of the customer, helping companies to better govern their data and support privacy and security initiatives. With the ability to better connect data across an organization, companies can more easily use AI and advanced analytics for real-time insights, "hydrate" business applications with critical data to make them more effective, and deliver a new category of AI-powered services for customers.

Another example is the Open Manufacturing Platform (OMP) which Microsoft and the BMW Group annouced at Hannover Messe in April 2019. The OMP is a new community initiative to enable faster, more cost-effective innovation in the manufacturing sector.

Through the creation of an open technology framework and cross-industry community, the OMP is expected to support the development of smart factory solutions that will be shared by OMP participants across the automotive and broader manufacturing sectors. The goal is to significantly accelerate future industrial IoT developments, shorten time to value and drive production efficiencies while addressing common industrial challenges.

Built on the Microsoft Azure industrial IoT cloud platform, the OMP is intended to provide community members with a reference architecture with open source components based on open industrial standards and an open data model. In addition to facilitating collaboration, this platform approach is designed to unlock and standardize data models that enable analytics and machine learning scenarios—data that has traditionally been managed in proprietary systems. Utilizing industrial use cases and sample code, community members and other partners will have the capability to develop their own services and solutions while maintaining control over their data.

With currently over 3,000 machines, robots and autonomous transport systems connected with the BMW Group IoT platform, which is built on Microsoft Azure's cloud, IoT and AI capabilities, the BMW Group plans to contribute relevant initial use cases to the OMP community. One example is the company's use of their IoT platform for the second generation of its autonomous transport systems in the BMW Group plant in Regensburg, Germany, one of 30 BMW Group production and assembly sites worldwide. This use case has enabled the BMW Group to greatly simplify its logistics processes via central coordination of the transport system, creating greater logistics efficiency. In the future, this and other use cases—such as digital feedback loops, digital supply chain management and predictive maintenance—will be made available and, in fact, developed further within the OMP community.

The ODI and OMP are just two examples of companies working together to derive better insights from diverse supply chain information. This can provide a model for other industries and companies to follow.

Capacity Building

Many business leaders and workers in the manufacturing sector believe that AI will have a positive impact on their jobs and organizations. This is a sentiment backed up by a recent study by IDC of business leaders and workers from across 15 countries in Asia Pacific.[26] The manufacturers who were interviewed spoke passionately about the urgency of expanding educational opportunities and skills development related to technology. The advent of AI has only increased this sense of urgency, and not only for technical skills. The business leaders interviewed identified the top three skills where demand will outstrip supply in the next three years: 1) communication and negotiation skills; 2) entrepreneurship and initiative-taking; and 3) adaptability and continuous learning.

At the same time, business leaders believe that demand for basic data processing, literacy and numeracy, and general equipment operations and mechanical skills will decrease in the next three years. Those surveys claimed those skills are broadly available today, and already now the supply is higher than the demand in Asia Pacific.

As Scott Hunter, Microsoft's Regional Business Lead for Manufacturing, commented, there is a disconnect between employers' perception of their workers' willingness to reskill and the actual willingness of workers.

"Business leaders are aware of the massive reskilling efforts required to build an AI-ready workforce. However, 22 percent of business leaders felt that workers have no interest in reskilling, but only 8 percent of workers feel the same. In addition, 48 percent of business leaders feel that workers do not have enough time to reskill, but only 34 percent of workers feel the same way."

What this means is that business leaders must prioritize reskilling and upskilling, dedicating employees' time for this to address the skills shortage. Even though it may result in short-term productivity impact, building an AI-ready workforce will result in greater gains in the future. For policy makers, a mobile and more dynamic workforce will increase pressure on the already constrained social safety net which will need to anticipate changes to unemployment insurance, reemployment programs – including job training and trade adjustment programs – and paid time for training.

With the customers we spoke to directly, some wanted to see support for "earn and learn" strategies. "We need a societal approach," said one.

"Readiness implies continued learning. We won't put people out of work," said another, "but we want to make sure they are trained to have the right job."

Another manufacturer was blunter: "We're going to displace people. People need to know that. We need some planning around that. What skills are needed? What is the next step?"

One executive recalled that the sector leaped onto the "Lean Manufacturing" bandwagon. Suddenly they realized people can lose jobs by focusing on waste reduction. "But they figured it out. I can't change people's mindset [about AI] if they fear it."

A Collaborative Approach

There was a consensus that, given the relative immaturity of AI, governments and policymakers need to avoid regulatory overreach, which so often can foreclose innovation opportunities. Manufacturers encouraged a more light-touch regulatory regime in the short term while governments continue to create a longer-term policy framework. As one manufacturer said, "AI is at a low level of intelligence today. We are expecting it to do things it can't. Its level of intelligence is like a mouse. AI is only as intelligent as we make it."

Another customer observed, "If government tries to control, we will find ourselves falling behind and miss the boat. In Europe, regulation is not empowering digitalization. Don't regulate for control but for benefits."

One executive noted that government does not yet understand AI in a way that would make their ideas constructive and helpful. "All governments are trying to catch up with the development of technologies. Legislation is always slower than the tech itself."

Beyond this light-touch regulatory approach, customers commented on the need to look at policy development from a whole-of-industry perspective and avoid making policy based on narrow considerations. Given the pervasive nature of AI across all industries, stakeholders are anxious that policymakers understand this intersectionality earlier and think about industry policy when they think about digital policy. "Do something that makes sense," one source cautioned. "Policymakers need to get to know the processes manufacturers use. Don't create impediments that would get in the way of connectivity of data. We are trying to improve process and make customers better, so don't get in the way of that."

"One big fear," said a senior executive at an American company, "is that we will get into a regulatory situation about data sharing and exchange. Bricks and mortar are no longer the asset of the company—data is. Government wants to protect, but with protection comes control."

There was also a strong view that governmental bodies and industry must come together to agree on sensible standards. There is also the need to think in strategic terms at both a national and an industry level. "The standards are not fully developed," one manufacturer told us. Having a rich set of AI standards will make it easier to scale training for workers and make it easier to deploy technology.

This was confirmed by a European company, which said that government needs to either embrace AI or be left behind. "Places with the highest concentration of robots have the lowest unemployment. Policy, academia, and industry need to come together."

One European policymaker stressed the need for harmonizing digital laws worldwide: "We can't adopt rules in our own corner and not talk with one another." Industry-led organizations such as the Partnership on AI,[27] which brings together industry, nonprofit organizations, and NGOs, can serve as forums for the process of devising best practices and entering into a dialogue with governments on the direction of future policy and regulation.

One place where these various policy threads can be woven together is through national AI strategies. These blueprints serve as a cross-sector and multi-stakeholder effort to align the various interests and needs of the public and private sectors as well as serve the broader interests of the community. To date, more than 20 countries have announced comprehensive national AI strategies, outlining how AI can boost national competitiveness while addressing societal concerns.[28]

Many of these strategies put manufacturing and Industry 4.0 as a key driver and have set up ethics panels and high-level expert groups to address the economic, ethical, policy, and legal implications of AI. The voice of business leaders, labor representatives, and technology providers will be critical to realizing the vision behind these plans.

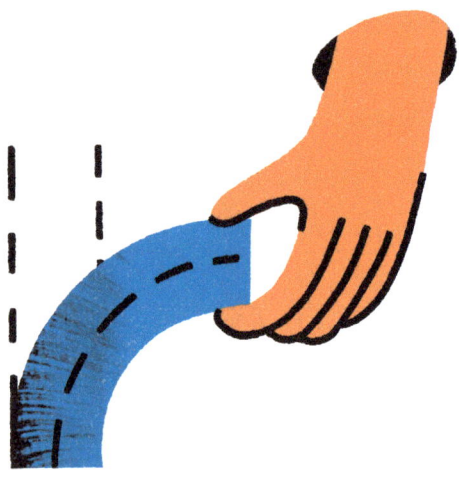

Environmental Sustainability

Manufacturing and production activities account for a significant impact on the environment. Global manufacturing consumes some 54 percent of the world's total energy and creates around one-fifth of global greenhouse gas emissions.[29] Further, the global water crisis is exacerbated by the almost insatiable need for water in the manufacturing of most goods, with millions of gallons of water being used each and every day to produce consumer goods. Just a few examples underscore this point: It takes between 13,737 to 21,926 gallons of water to make a car; 3,626 gallons for a pair of leather shoes; 3,190 gallons for a smartphone; 2,108 gallons for a pair of jeans; and 659 gallons for a T-shirt.[30]

This growing water and energy consumption in manufacturing is set against an already challenging environmental situation: communities that are ravaged by floods and wildfires; farmers who are losing their harvests to pests and extreme weather; warming oceans, deforested wilderness, arid soil, and contaminated watersheds. As these crises unfold, they are devastating for local communities who go hungry, thirsty, or lose their homes and livelihoods. Environmental challenges also impact the broader ecosystem, rippling across national boundaries and disrupting the global economy, costing billions of dollars in business and property damage, not to mention loss of human life.[31]

There is a globally recognized imperative to become more sustainable and address the world's most pressing environmental challenges. Scientists say we have less than two decades to course correct the impact humans are having on the planet,[32] and United Nations member states issued an urgent call for action and a global partnership to tackle this, including promoting responsible production.[33]

FOSTERING RESPONSIBLE INNOVATION

The challenge is clear—manufacturers must break the pattern of past industrial revolutions, and deliver not just economic growth, but sustainable growth.

Sustainability also has significant reputational value. A report by McKinsey & Company[34] notes that environmental sustainability is now a strategic and integral part of many businesses. Indeed, a Nielsen study of 30,000 consumers across the world indicates that 73 percent of Millennials are willing to pay extra for sustainable goods.[35]

Therefore, sustainable manufacturing is no longer optional—it's a business imperative. Manufacturers need to leverage the best technologies possible to harness vast amount of data and make breakthrough advances in reducing environmental impact and managing scarce resources.

Manufacturers have already started leveraging AI for sustainable production, ensuring it is economically and environmentally sustainable. For example, companies such as Ecolab, Ørsted, and The Yield have developed AI-enabled solutions to improve water conservation, renewable energy management, and agricultural production. A small startup, SilviaTerra, is using AI to improve forest management and fuel the carbon offset market with precise data about the carbon sequestration potential of every acre of forestland in the United States.

These companies' experiences demonstrate that AI can enable innovations in sustainable manufacturing and go hand in hand with profit-making and improved competitiveness.

These are not isolated examples. New research from PwC shows that AI can be applied to a wide range of economic sectors and industries to improve environmental outcomes and mitigate climate change—all while driving economic growth.[36] In fact, in their research, PwC estimates that using AI for environmental applications could contribute up to US $5.2 trillion to the global economy in 2030, a 4.4 percent increase relative to business as usual, and create an estimated 38.2 million new skilled, green jobs across the global economy. The report highlights how AI is increasingly part of the suite of options in heavy industry, which alongside heavy transport, has often been called out as "hard to abate."

Additional examples include:

- **Cement:** AI, often combined with advanced sensors, is being used for predictive asset management to maximize the efficiency, operation, and management of production assets.

- **Steel:** AI systems are increasingly being tried and tested in process and operational controls to supplement traditional controls, increase efficiency, and optimize system operations.

- **Chemicals:** AI, often combined with traditional controls and new sensors, is being deployed for predictive maintenance in chemicals manufacturing, maximizing efficiency and minimizing resource use in chemicals processes.

- **Shipping/supply chain:** AI can be used in the maritime sector a range of ways, including for predictive vessel management and maintenance, real-time voyage optimization, and fuel monitoring and management.[37]

And with the democratization of AI, small and medium-sized business now have access to these same innovative technologies and are starting to embrace these great opportunities.

They are learning from the earlier pioneers and are leveraging resources such as the "OECD Sustainable Manufacturing Toolkit", which aims to provide a practical starting point for businesses around the world to improve the efficiency of their production processes and products, enabling them to contribute to sustainable development and green growth.[38] The toolkit includes an internationally applicable common set of indicators to help businesses measure their environmental performance at the level of a plant or facility.[39]

There is enormous potential for AI to be an important tool in the effort to decouple economic growth from rising carbon emissions. There is a clear path for manufacturers toward a prosperous and more sustainable future with advanced technologies.

This is the future of sustainable manufacturing and environmental sustainability for our planet, and AI has a substantial role to play. Whether it's deployed to monitor and improve environmental outcomes, enhance resource management, or create a greener economy, taking a tech-first, AI-enabled approach is the path forward.

Where to from Here

As with the great advances of the past on which it builds—from the steam engine, to the combustion engine, from electricity to the microprocessor—AI will bring about vast changes to the manufacturing sector, some of which are already being realized by organizations today.

But just as with these previous significant technological advances, we'll need to be thoughtful about how we address the economic, employment, societal, and environmental issues that these changes bring about. Most importantly, we all need to work together—industry, government, academia, educators, civil society, and labor representatives—to ensure that AI is developed and deployed in a responsible and trustworthy manner.

Each of us has a responsibility to participate and an important role to play. For our part, one of those roles is helping organizations of all sizes realize their tech intensity vision. The next chapter is all about helping customers on their AI journey and their pathway to innovation.

AI AND THE CONNECTED SUPPLY CHAIN: THE MICROSOFT EXPERIENCE

Sitting in her office on the Microsoft campus, Pat Flynn-Cherenzia glances at her computer screens and quickly assesses the company's supply chain all over the world. At the moment, shipments of Microsoft supplies and inventory look mostly green along the U.S. coasts. Normally at this time of year she can see problems in Florida. It's red in Europe due to ice, and outbound traffic in Asia is yellow as businesses rush to get product out of ports before real and threatened tariffs hit.

Pat uses AI systems to help predict when certain events will occur. She's led logistics at Microsoft for years, but earlier in her career she worked for U.S. Customs, the ultimate supply chain.

"So much is coming and going, you couldn't inspect everything so you had to combine datasets and create algorithms that could tell you, 'this is a viable candidate for inspection or this is a trusted source.'"

A native of Philadelphia, Pat worked for U.S. Customs there, where her goal was to prevent intrusions at that city's port. She was on the lookout for first-time importers from places where the threat was elevated and the shipper had a "funky product," like say a rickshaw from Thailand. Turns out that shipment was full of narcotics, and she caught it. Fast forward to today, and Pat is a key player in Microsoft's own digital transformation in manufacturing. She oversaw multiple "control towers" for Nokia's factories and distribution centers, speeding up the onboarding of new suppliers like DHL by moving to a single platform in the cloud.

Later, when a malware attack hit the shipping giant Maersk, it decimated their systems, causing the shipper and its customers to

lose visibility. But with Microsoft's cloud and cognitive services in place, Pat was able to use independent data to triangulate and predict where shipments were within 20 minutes after the attack.

"We were able to tell them where their shipments were so they could be retrieved."

Moving to the cloud has allowed Microsoft to leverage machine learning and predictive analytics, which lets Pat anticipate disruptions. At the time of the Xbox launch, South Korea's Hanjin Shipping was offering attractive rates, but the firm abruptly filed for bankruptcy and it took a week to find Microsoft's inventory. "Even if the provider doesn't tell me, I know by SKU where my products are on boats. Satellites are tracking my goods. Based on trends, I can predict when the product will arrive at port. If I see a weather pattern building in the Caribbean, and I see our product is headed in that direction, I can hold it back."

During Hurricane Michael in October 2018, Pat held back shipments headed for the Florida panhandle and instead diverted to Memphis just in time. Microsoft gives the same visibility to its customers, so in this case the retailer Best Buy could know with confidence when it would see its product.

Pat has since moved onto another role within Microsoft to share the depth and breadth of her experiences with small, medium and large manufacturing customers worldwide.

CHAPTER 4

THE WAY
FORWARD

THE WAY FORWARD

AI MATURITY AND PROGRESSING YOUR JOURNEY

As we have seen, AI is a critical component in helping manufacturing organizations transform their operations, better serve customers, and offer new opportunities to their workers. To realize this potential, each organization needs to adopt a strategy, culture, and set of core capabilities around AI that match the organization's maturity. The following section looks at how organizations can progress their AI journey in a way that corresponds to their maturity level.

> **AI-based systems must be continuously trained, monitored, and evaluated for performance if organizations are to realize their full benefits while guarding against bias, privacy violations, and safety concerns**

Deploying an AI-based system is very different than acquiring traditional packaged software or developing a custom-coded non-AI solution. AI-based systems must be continuously trained, monitored, and evaluated for performance if organizations are to realize their full benefits while guarding against bias, privacy violations, and safety concerns. Neglecting this maturity assessment can seriously impede an AI project, potentially causing a rejection of the technology by employees who perceive it as too difficult to use or untrustworthy.

To address these challenges, Microsoft has worked with customers in all sectors of the economy to define an operational model that helps organizations assess their own attributes and maturity. Microsoft's AI Maturity Model helps organizations conduct this assessment and guides adoption of the right kind of AI at the right place and time, smoothing the path to safe deployment of progressively more advanced capabilities.

The AI Maturity Model includes four stages of development:

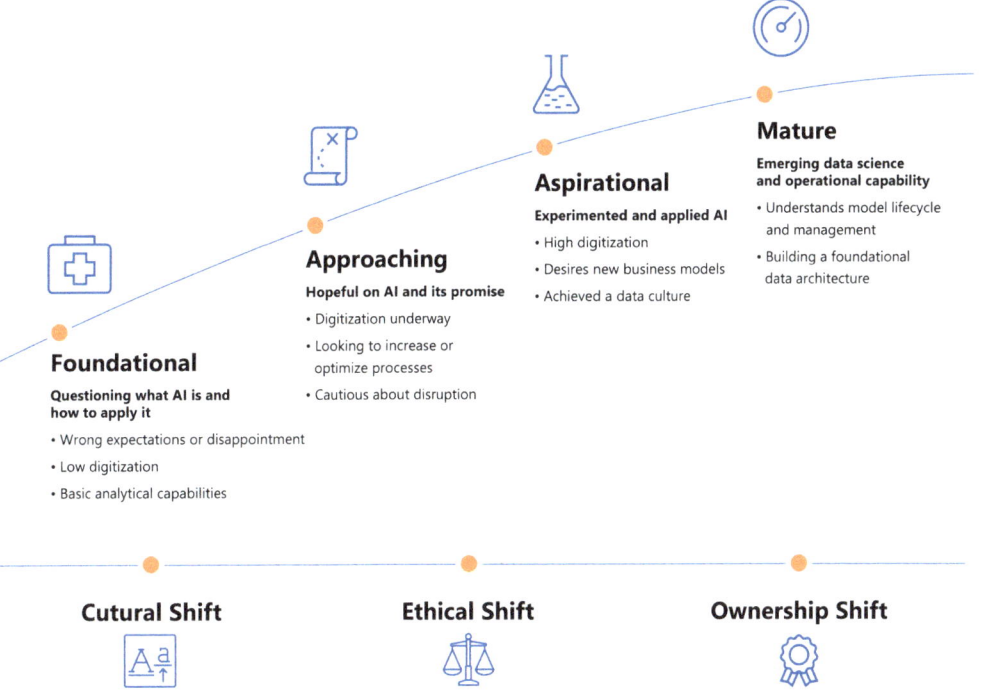

Foundational

Organizations at the foundational stage are seeking to understand the varieties and applications of AI and how others in their industry are using it.

They strive to make more data-driven decisions and currently tend to rely on the instincts of experienced leaders to make decisions.

Foundational organizations need to invest in projects that focus on fast, iterative experimentation. Doing this successfully requires organizations to build a culture that embraces experimentation and empowers colleagues to make data-driven decisions. Organizations at this level of maturity should look to adopt AI technologies built on established platforms, helping them to grow into digital businesses.

Approaching

Organizations at this level of maturity are implementing cultural changes to empower employees and make data-driven decisions. They are focused on building a data culture and using AI to build new processes and streamline operations. Having digitized assets and deployed AI to automate certain processes, these organizations are ready to explore owning custom AI solutions. Approaching organizations are poised to embrace rapid experimentation and will invest more in understanding how to implement and improve AI over time.

Investments should continue in accountability protocols for AI governance, monitoring, orchestrating, and improving AI over time and infusing ethical viewpoints in AI-based systems. Considering these issues will help organizations gain experience when using AI to digitally transform.

Aspirational

Aspirational organizations understand that AI will be instrumental in helping them compete and transform. These organizations know that others are using AI and understand the competitive disruption this may entail.

Organizations at this maturity level are focused on shifting culture to empower employees, increasing collaboration, and generating ideas for optimization, new offerings, and business models. These organizations are becoming increasingly comfortable with taking risks and are striving to transition away from fixed projects to more iterative projects.

Aspirational organizations can adopt configurable AI, hosted by technology companies. This abstracts away the operational complexity of maintaining the core AI while allowing organizations to infuse AI into digital experiences. At the same time, experimentation with more advanced AI technologies such as custom AI is encouraged to these organizations learn about how to operate and coordinate more complex systems.

Mature

Mature organizations have shifted their culture to embrace rapid, iterative experimentation and a data-driven approach.

Mature organizations develop AI talent and understand how to apply this resource to multiple AI initiatives simultaneously.

Mature organizations ask not just what can we do with AI but what should we do with AI. The organizations also infuse ethical perspectives into their experience creation process.

> **Mature organizations ask not just what can we do with AI but what should we do with AI. The organizations also infuse ethical perspectives into their experience creation process.**

Organizations at this level of maturity should continue to evaluate tool chains for configurable and custom AI while being vigilant about monitoring, retraining, and updating AI-based systems. Maintaining AI talent, prioritizing new strategic initiatives, and continued agile experimentation are areas of focus for mature organizations.

A Path to Innovation

Digital transformation is a journey. It provides a path for organizations to innovate and establish new and better ways of doing business that benefit all stakeholders.

Considering and assessing an organization's AI maturity provides a clear path to guide AI technology adoption efforts. Organizations closer to the Foundational or Approaching stages should look to adopt configuration-based AI first, where the concepts of operationalization are developed by partners like Microsoft. When more specific AI capabilities are required, organizations should assess themselves across strategic, cultural, and capability boundaries to determine if they are ready to own and operate a custom AI solution.

Microsoft works with customers to prepare, host, and implement configurable and custom AI to help organizations digitally transform. Through a combination of AI services, AI platform capabilities, and smart digital experiences, we work to make the infusion of AI into digital experiences a rapid, iterative process that speeds transformation.

Enterprise leaders should recognize that we are all transitioning from an era when every company is a digital company to one where every company must now become an AI company. And manufacturers now need to be thinking as digital and AI companies. This will be a cultural shift, as well as a technical and business one.

CONCLUSION

As this book went to press, more than 200,000 manufacturers and 6,000 exhibiting companies converged on the German city of Hannover for the annual tradeshow, Hannover Messe.

On display at the tradeshow were the very latest products and trends in industrial technology. Mile after crowded mile, in cavernous halls the size of airplane hangars, manufacturers from every corner of the globe were treated to stunning displays of ever-evolving automation—robots, sensors, IoT breakthroughs, AI and machine learning, drones, and every manner of drill, assembly line, and factory floor machine.

It's clear that Industry 4.0 has arrived. Manufacturers are now digital companies.

While the machinery on show was impressive, it's what you can't see that takes center stage—data. It's clear that Industry 4.0 has arrived. Manufacturers are now digital companies. Not surprisingly, the effect was inspiring and just a little overwhelming. Not unlike the premise of this book.

When we set out to write this book, we had one simple objective: listen to manufacturing customers and public policy experts to learn about their AI journey, report their stories, and offer our own insights and recommendations on the implications and possibilities that lie ahead.

At Hannover Messe, one of our customer and partners, Schneider Electric, was on hand to demonstrate how its platform is leveraging Microsoft AI to help a range of customers stay ahead of maintenance problems in applications as varied as coffee roasters in the developed world to schools and clinics in the developing world.

One example was their use of Microsoft's Cortana Intelligence Suite in Nigeria, where Schneider can identify trends in its solar panels so technicians can address issues before they lead to outages.

Historical data might show that a certain drop in electricity generated by a solar panel may indicate that a panel needs to be cleaned or a battery checked within 12 hours or it could fail. The analytics allow remote monitors to help proactively ward off those types of problems.

Another of our customers at Hannover Messe was Repsol, one of the world's largest oil and gas exploration and production companies. They spoke to us about how AI is helping to fuel the world's energy demands while also protecting the environment. Data and computing make it possible to see beneath layers of soil, sequence DNA, and therefore make drilling more precise; minimizing not just the number of wells needing to be drilled, but maximizing the time to drill them. All of this helps reduce the time from acquisition of the site to when the energy is commercialized.

Every manufacturer with whom we engaged reported that AI is boosting efficiency, productivity, safety, and health. Those policymakers we spoke to also reported that a stronger economy from a more efficient, sustainable, and competitive manufacturing sector will lead to improvements for their constituents and for the economy overall. The economic impact was not the only issue on which manufacturers and policymakers agreed; they also were in agreement that an AI-led future requires a commitment to bringing along the current workforce and cultivating a long-term skills pipeline for future generations.

To underscore the implications for the next generation of talent, one executive told a story about a recent tour of colleges for prospective students where a career counselor asked the applicants what year they would graduate high school and then what year they would graduate college. Students and parents alike were confident, and the answers came fast and furious. But when asked what year they would retire, the audience fell silent. The counselor then estimated that, whenever their retirement might be, the students would have worked in seven different fields, two of which had not yet been invented. In other words, there will be disruption, and we must all prepare.

Fortunately, that is happening. According to LinkedIn Chief Economist Guy Berger, the presence of AI-related certifications in American manufacturing is higher than in the typical industry and rising, with Guy reporting that 0.86 percent of certifications in U.S. manufacturing are AI related as of 2018. This is more than double the share seen two years ago. While still far below leading-edge tech industries like software, IT services, hardware, and networking, manufacturing ranks seventh out of 22 industries and scores higher than the median, or typical, industry.

In the executive summary of this book we note six findings from our inquiry. In the customer stories we highlight a range of takeaways. In the skills section we speak about the urgency to address the changing workplace and workforce. And in the policy section we outline ethical and regulatory considerations.

Throughout your reading of this book, we hope that a single overarching theme emerges; one of optimism. Optimism that the inevitable march of technological progress can, as it has for centuries, lead to better, safer, and healthier lives when we work together to harness innovation for good and plan for the unforeseen consequences with shared legal and ethical principles.

We hope you will continue the discussion with your questions, observations, and recommendations by joining us at:
https://news.microsoft.com/futurecomputed

CONCLUSION

FURTHER READING

Accenture (website). "Industry X.O." Accessed March 2019. https://www.accenture.com/us-en/insights/industry-x-0-index.

Accenture Research. "Pivoting with AI." Accenture, 2018. https://www.accenture.com/t20180912T132343Z__w__/za-en/_acnmedia/PDF-85/Accenture-Pivoting-with-AI-POV.pdf.

Acemoglu, Daron and Pascual Restrepo. "Artificial Intelligence, Automation, and Work." In: The Economics of Artificial Intelligence: An Agenda, ed. Ajay Agrawal, Joshua Gans, and Avi Goldfarb. Cambridge, MA: National Bureau of Economic Research, May 2019. p. 197-236. https://www.nber.org/chapters/c14027.

Bauer, Harald, Peter Breuer, Gérard Richter, Jan Wüllenweber, Knut Alicke, Matthias Breunig, Ondrej Burkacky, et al. "Smartening up with Artificial Intelligence (AI) – What's in it for Germany and its Industrial Sector?" McKinsey & Company, April 2017. https://www.mckinsey.com/~/media/McKinsey/Industries/Semiconductors/Our%20Insights/Smartening%20up%20with%20artificial%20intelligence/Smartening-up-with-artificial-intelligence.ashx.

Atluri, Venkat, Saloni Sahni, and Satya Rao. "The trillion-dollar opportunity for the industrial sector: How to extract full value from technology." McKinsey Digital, November 2018. https://www.mckinsey.com/business-functions/digital-mckinsey/our-insights/the-trillion-dollar-opportunity-for-the-industrial-sector.

Castellina, Nicholas. "How Artificial Intelligence is Transforming the Manufacturing Workforce." Manufacturing.net, September 2018. https://www.manufacturing.net/article/2018/09/how-artificial-intelligence-transforming-manufacturing-workforce.

Charalambous, Eleftherios, Robert Feldmann, Gérard Richter, and Christoph Schmitz. "AI in production: A game changer for manufacturers with heavy assets." McKinsey & Company, March 2019. https://www.mckinsey.com/business-functions/mckinsey-analytics/our-insights/ai-in-production-a-game-changer-for-manufacturers-with-heavy-assets.

Chui, Michael and Sankalp Malhotra. "AI adoption advances, but foundational barriers remain." McKinsey & Company, November 2018. https://www.mckinsey.com/featured-insights/artificial-intelligence/ai-adoption-advances-but-foundational-barriers-remain.

Coleman, Chris, Satish Damodaran, Mahesh Chandramouli, and Ed Deuel. "Making Maintenance Smarter: Predictive Maintenance and the Digital Supply Network." Deloitte Insights, May 9, 2017.

https://www2.deloitte.com/insights/us/en/focus/industry-4-0/using-predictive-technologies-for-asset-maintenance.html.

Forbes Insights. "How AI Builds A Better Manufacturing Process." Forbes, July 17, 2018. https://www.forbes.com/sites/insights-intelai/2018/07/17/how-ai-builds-a-better-manufacturing-process/#28714b391e84.

Geissbauer, Reinhard, Evelyn Lübben, Stefan Schrauf, and Steve Pillsbury. Global Digital Operations 2018 Survey. PwC Strategy&, 2018. https://www.strategyand.pwc.com/media/file/Global-Digital-Operations-Study_Digital-Champions.pdf.

Giffi, Craig, Paul Wellener, Ben Dollar, Heather Ashton Manolian, Luke Monck, and Chad Moutray. "2018 Skills Gap in Manufacturing Study." Deloitte Insights, 2018. https://www2.deloitte.com/us/en/pages/manufacturing/articles/future-of-manufacturing-skills-gap-study.html.

Goering, Kevin, Richard Kelly, and Nick Mellors. "The Next Horizon for Industrial Manufacturing: Adopting Disruptive Digital Technologies in Making and Delivering." McKinsey Digital, November 2018. https://www.mckinsey.com/business-functions/digital-mckinsey/our-insights/the-next-horizon-for-industrial-manufacturing.

Hawksworth, John and Yuval Fertig. "What Will Be the Net Impact of AI and Related Technologies on Jobs in China?" PwC, September 2018. https://www.pwc.com/gx/en/issues/artificial-intelligence/impact-of-ai-on-jobs-in-china.pdf.

International Federation of Robotics. "Robot Density Rises Globally." International Federation of Robotics press release, February 7, 2018. https://ifr.org/ifr-press-releases/news/robot-density-rises-globally.

Küpper, Daniel, Markus Lorenz, Kristian Kuhlmann, Olivier Bouffault, Yew Heng Lim, Jonathan Van Wyck, Sebastian Köcher, et al. "AI in the Factory of the Future: The Ghost in the Machine." Boston Consulting Group, April 2018. https://www.bcg.com/en-us/publications/2018/artificial-intelligence-factory-future.aspx.

Kushmaro, Philip. "5 Ways Industrial AI is Revolutionizing Manufacturing." CIO, September 27, 2018. https://www.cio.com/article/3309058/5-ways-industrial-ai-is-revolutionizing-manufacturing.html.

Microsoft Library ProResearch. "Artificial Intelligence (AI) & Ethics." Topic paper, Microsoft Library ProResearch, June 11, 2018. https://microsoft.sharepoint.com/sites/mslibrary/KeyTopics/Pages/Research/AI%20Ethics.pdf.

Mittal, Nitin, David Kuder, and Samir Hans. "AI-Fueled Organizations: Reaching AI's Full Potential in the Enterprise." Deloitte Insights, January 16, 2019. https://www2.deloitte.com/insights/us/en/focus/tech-trends/2019/driving-ai-potential-organizations.html.

Muro, Mark, Robert Maxim, and Jacob Whiton. Automation and Artificial Intelligence: How Machines are Affecting People and Place. With contributions by Ian Hathaway. Washington, D.C.: Metropolitan Policy Program at Brookings, January 24, 2019. https://www.brookings.edu/wp-content/uploads/2019/01/2019.01_BrookingsMetro_Automation-AI_Report_Muro-Maxim-Whiton-FINAL-version.pdf.

Perisic, Igor. "How Artificial Intelligence is Already Impacting Today's Jobs." Economic Graph (blog), LinkedIn, September 17, 2018. https://economicgraph.linkedin.com/blog/how-artificial-intelligence-is-already-impacting-todays-jobs.

Philbeck, Thomas, Nicholas Davis, and Anne Marie Engtoft Larsen. "Values, Ethics and Innovation Rethinking Technological Development in the Fourth Industrial Revolution." Cologny-Geneva: World Economic Forum, January 2018. http://www3.weforum.org/docs/WEF_WP_Values_Ethics_Innovation_2018.pdf.

PwC. "AI Will Create as Many Jobs as it Displaces by Boosting Economic Growth." PwC press release, July 17, 2018. https://www.pwc.co.uk/press-room/press-releases/AI-will-create-as-many-jobs-as-it-displaces-by-boosting-economic-growth.html.

Rajadhyaksha, Ajay and Aroop Chatterjee. "Robots at the gate: Humans and technology at work." Barclays, April 11, 2018. https://www.investmentbank.barclays.com/content/dam/barclaysmicrosites/ibpublic/documents/our-insights/Robots-at-the-gate/Barclays-Impact-Series-3-Robots_at_the_Gate-3MB.pdf.

Ransbotham, Sam, Philipp Gerbert, Martin Reeves, David Kiron, and Michael Spira. "Artificial Intelligence in Business Gets Real Pioneering Companies Aim for AI at Scale." MIT Sloan Management Review (Access notes), September 2018. https://sloanreview.mit.edu/projects/artificial-intelligence-in-business-gets-real/.

Rao, Dr. Anand S. and Gerard Verweij. "Sizing the prize: What's the real value of AI for your business and how can you capitalise?" PwC, 2017. https://www.pwc.com/gx/en/issues/analytics/assets/pwc-ai-analysis-sizing-the-prize-report.pdf.

Rebello, Jagdish and Sachin Garg. Artificial Intelligence in Manufacturing Market – Global Forecast to 2023. With contributions by Tanuj Goyal, Siddharth Rane, Anand Shanker, and Tarun Dutta. MarketsandMarkets, July 2017. https://microsoft.sharepoint.com/sites/mslibrary/research/ MktResearch/MarketsAndMarkets/Artificial_Intelligence_in_ Manufacturing_Market_Global_Forecast.pdf#search=artificial%20 intelligence%20manufacturing.

Rodriguez, Michelle Drew, Robert Libbey, Sandeepan Mondal, Jeff Carbeck, and Joann Michalik. "Exponential Technologies in Manufacturing." Deloitte, 2018. https://www2.deloitte.com/us/en/pages/ manufacturing/articles/advanced-manufacturing-technologies-report. html.

Rose, Justin, Vladimir Lukic, Claudio Knizek, Tom Milton, Alex Melecki, and Howie Choset. "Advancing Robotics to Boost US Manufacturing Competitiveness." Boston Consulting Group, October 25, 2018. https:// www.bcg.com/en-us/publications/2018/advancing-robotics-boost-us- manufacturing-competitiveness.aspx.

Schaeffer, Eric, Jean Cabanes, and Abhishek Gupta. "Manufacturing the Future." Accenture, 2018. https://www.accenture.com/ t20180327T080053Z__w__/us-en/_acnmedia/PDF-74/Accenture-Pov-Manufacturing-Digital-Final.pdf#zoom=50.

Schaeffer, Eric, Raghav M. Narsalay, Oliver Hobräck, Matthias Wahrendorff, and Abhishek Gupta. "Turning Possibility into Productivity." Accenture, 2018. https://www.accenture.com/ t20180516T150528Z__w__/us-en/_acnmedia/PDF-78/Accenture-IndustryX0-AI-products_RD.pdf#zoom=50.

Shaw, Jonathan. "Artificial Intelligence and Ethics." Harvard Magazine, January-February 2019. https://harvardmagazine.com/2019/01/artificial-intelligence-limitations.

Shum, Harry. "How to Ensure that We're 'Raising' Ethical AI." LinkedIn Pulse (blog), LinkedIn, May 17, 2018. https://www.linkedin.com/pulse/how-ensure-were-raising-ethical-ai-harry-shum/.

Somers, Ken. "Manufacturing's Control Shift." McKinsey & Company, August 22, 2018. https://www.mckinsey.com/business-functions/operations/our-insights/operations-blog/manufacturings-control-shift.

The Manufacturer. "Annual Manufacturing Report 2018." London: Hennik Research, May 2018. https://cdn2.hubspot.net/hubfs/728233/AMR-2018-Amended-May-2018.pdf.

Tratz-Ryan, Bettina, Alexander Hoeppe, and Pete Basiliere. "Predicts 2018: Industries 4.0 and Advanced Manufacturing." Gartner (Access notes), April 9, 2018. https://www.gartner.com/document/3871170.

Tratz-Ryan, Bettina and Michelle Duerst. Manufacturing Industries Digitalization Primer for 2019. Gartner (Access notes), February 5, 2019. https://www.gartner.com/document/3899977.

Wellener, Paul, Ben Dollar, and Heather Ashton Manolian. "The Future of Work in Manufacturing." Deloitte Insights, March 21, 2019. https://www2.deloitte.com/insights/us/en/industry/manufacturing/future-of-work-manufacturing-jobs-in-digital-era.html.

World Economic Forum and McKinsey & Company. "Fourth Industrial Revolution: Beacons of Technology and Innovation in Manufacturing." Cologny-Geneva: World Economic Forum, January 2019.

http://www3.weforum.org/docs/WEF_4IR_Beacons_of_Technology_and_Innovation_in_Manufacturing_report_2019.pdf.

World Economic Forum and McKinsey & Company. "The Next Economic Growth Engine: Scaling Fourth Industrial Revolution Technologies in Production." Cologny-Geneva: World Economic Forum, January 2018. http://www3.weforum.org/docs/WEF_Technology_and_Innovation_The_Next_Economic_Growth_Engine.pdf.

Zhong, Ray Y., Xun Xu, Eberhard Klotz, and Stephen T. Newman. "Intelligent Manufacturing in the Context of Industry 4.0: A Review." Engineering vol. 3, no. 5 (October 2017): 616-630. https://doi.org/10.1016/J.ENG.2017.05.015.

FURTHER READING

ENDNOTES

1   Long, Katie, Donna Tam, Carrie Barber, Arjuna Soriano, and Paul Brent. "Robot-Proof Jobs." Marketplace, 2017. https://features.marketplace.org/robotproof/

2   Prodhan, Georgina. "European parliament calls for robot law, rejects robot tax." Reuters Technology News, Feb. 16, 2017. https://www.reuters.com/article/us-europe-robots-lawmaking-idUSKBN15V2KM

3   Giffi, Craig A., Paul Wellener, Ben Dollar, Heather Ashton Manolian, Luke Monck, and Chad Moutray. "2018 skills gap in manufacturing study, Future of manufacturing: The jobs are here, but where are the people?" Deloitte, Nov. 14, 2018. https://www2.deloitte.com/us/en/pages/manufacturing/articles/future-of-manufacturing-skills-gap-study.html

4   "Which Roles are Most Difficult to Fill in the UK?" 2018 Talent Shortage Survey. Manpower Group, Feb. 7, 2018. https://www.manpowergroup.co.uk/the-word-on-work/2018-talent-shortage-survey/#hard-to-fill-roles

5   Rao, Dr. Anand S., and Gerard Verweij. Sizing the prize: What's the real value of AI for your business and how can you capitalise? PwC, 2017. https://www.pwc.com/gx/en/issues/analytics/assets/pwc-ai-analysis-sizing-the-prize-report.pdf

6   Purdy, Mark, and Paul Daugherty. "Future of Artificial Intelligence: Artificial Intelligence Is The Future Of Growth." Accenture LLP. Accessed May 19, 2019. https://www.accenture.com/us-en/insight-artificial-intelligence-future-growth

7   "The Federal Government's Artificial Intelligence Strategy: 'AI Made in Germany." German Federal Ministry for Economic Affairs and Energy. Accessed May 19, 2019. https://www.de.digital/DIGITAL/Redaktion/EN/Standardartikel/artificial-intelligence-strategy.html

8   Gates, Dominic. "Boeing invests in advanced supersonic business jet." Seattle Times, Feb. 5, 2019. https://www.seattletimes.com/business/boeing-aerospace/boeing-invests-in-advanced-supersonic-business-jet/

9   "Will AI Destroy More Jobs Than It Creates Over the Next Decade?" Wall Street Journal, April 1, 2019. https://www.wsj.com/articles/will-ai-destroy-more-jobs-than-it-creates-over-the-next-decade-11554156299

10   Greenhouse, Steven. "Unions Face The Fight Of Their Lives To Protect American Workers." Huffington Post, May 4, 2018. https://www.huffpost.com/entry/american-workers-jobs-inequality-union-automation_n_5ae043f9e4b061c0bfa32e0c

11   "UNI Global Union calls for the establishment of a global convention on ethical artificial intelligence." UNI Global Union, Dec. 14, 2016. https://www.uniglobalunion.org/news/uni-global-union-calls-establishment-a-global-convention-ethical-artificial-intelligence

12   "10 Principles for Ethical AI." UNI Global Union. Accessed May 19, 2019. http://www.thefutureworldofwork.org/opinions/10-principles-for-ethical-ai/

13   Meeting of the OECD Global Parliamentary Network, February 13-15, 2019. http://www.oecd.org/parliamentarians/meetings/gpn-meeting-february-2019/

14   "2017 Survey Of Occupational Injuries & Illnesses: Charts Package." U.S. Bureau of Labor Statistics. U.S. Department of Labor, Nov. 8, 2018. https://www.bls.gov/iif/osch0062.pdf

15   LinkedIn Workforce Report. Accessed May 19, 2019. https://economicgraph.linkedin.com/resources#view-all

16   Perisic, Igor. "How artificial intelligence is already impacting today's jobs." LinkedIn Economic Graph, Sept. 17, 2018. https://economicgraph.linkedin.com/blog/how-artificial-intelligence-is-already-impacting-todays-jobs

17	"The skills gap in U.S. manufacturing 2015 and beyond." Manufacturing Institute and Deloitte, 2015. http://www.themanufacturinginstitute.org/~/media/827DBC76533942679A15EF7067A704CD.ashx

18	"Manufacturing Our Future: Cases on the Future of Manufacturing." World Economic Forum, May 2016. http://www3.weforum.org/docs/GAC16_The_Future_of_Manufacturing_report.pdf

19	Boyle, Alan. "Microsoft CEO Satya Nadella lays out 10 Laws of AI (and Human Behavior)." Geekwire, June 28, 2016. https://www.geekwire.com/2016/microsoft-ceo-satya-nadella-10-laws-ai/

20	Wang, Amy B. "'I'm in your baby's room': A hacker took over a baby monitor and broadcast threats, parents say." Washington Post, Dec. 20, 2018. https://www.washingtonpost.com/technology/2018/12/20/nest-cam-baby-monitor-hacked-kidnap-threat-came-device-parents-say/

21	Content of Premarket Submissions for Management of Cybersecurity in Medical Devices: Draft Guidance for Industry and Food and Drug Administration Staff. U.S. Food and Drug Administration, Oct. 18, 2018. https://www.fda.gov/media/119933/download

22	"Age Demographics for Industry Workforces." Governing. Accessed May 19, 2019. https://www.governing.com/gov-data/ages-of-workforce-for-industries-average-medians.html

23	Romei, Valentina. "How Japan's ageing population is shrinking GDP." Financial Times, May 16, 2018. https://www.ft.com/content/7ce47bd0-545f-11e8-b3ee-41e0209208ec

24	"Shaw, Greg. The Ability Hacks. Microsoft, 2018. https://blogs.microsoft.com/wp-content/uploads/prod/sites/5/2018/08/theabilityhacksbook.pdf

25   "Adobe, Microsoft and SAP announce the Open Data Initiative to empower a new generation of customer experiences." Microsoft News Center, Sept. 24, 2018. https://news.microsoft.com/2018/09/24/adobe-microsoft-and-sap-announce-the-open-data-initiative-to-empower-a-new-generation-of-customer-experiences/

26   "Manufacturers with artificial intelligence to nearly double competitiveness." Microsoft Asia News Center, April 1, 2019. https://news.microsoft.com/apac/2019/04/01/manufacturers-with-artificial-intelligence-to-nearly-double-competitiveness/

27   Partnership on AI. Accessed May 19, 2019. https://www.partnershiponai.org/

28   "National And International AI Strategies." Future of Life Institute. Accessed May 19, 2019. https://futureoflife.org/national-international-ai-strategies/?cn-reloaded=1

29   Accelerating Sustainable Production. World Economic Forum. Accessed May 19, 2019. https://www.weforum.org/projects/accelerating-sustainable-production

30   "The Hidden Water In Everyday Products." Water Footprint Calculator, July 1, 2017. https://www.watercalculator.org/water-use/the-hidden-water-in-everyday-products/

31   "The Hidden Water In Everyday Products." FirstPost, April 22, 2019. https://www.firstpost.com/tech/news-analysis/earth-day-2019-ai-has-a-huge-role-to-play-in-fulfilling-sustainable-development-goals-6486441.html

32   "Global Warming of 1.5 °C." Intergovernmental Panel on Climate Change. Accessed May 19, 2019. https://www.ipcc.ch/sr15/

33   United Nations Sustainable Development Goals. United Nations. Accessed May 19, 2019. https://sustainabledevelopment.un.org/sdgs. See UN Sustainable Development Goals 12 and 13.

34   "Sustainability's Strategic Worth." McKinsey & Co., July 2014. https://www.mckinsey.com/business-functions/sustainability/our-insights/sustainabilitys-strategic-worth-mckinsey-global-survey-results

35   Banham, Russ. "Industry 2050: How Clean Manufacturing Is A Win-Win Proposition." Mitsubishi Heavy Industries Brandvoice. Forbes, Oct. 18, 2018, https://www.forbes.com/sites/mitsubishiheavyindustries/2018/10/18/industry-2050-how-clean-manufacturing-is-a-win-win-proposition/#473a13b17f35

36   Herweijer, Celine, Benjamin Combes, Jonathan Gillham, Lucas Joppa, et al. "How AI Can Enable a Sustainable Future." Microsoft and PwC. Accessed May 19, 2019. https://www.pwc.co.uk/sustainability-climate-change/assets/pdf/how-ai-can-enable-a-sustainable-future.pdf

37   "How AI can enable a sustainable future: Estimating the economic and emissions impact of AI adoption in agriculture, water, energy and transport." PwC UK. Accessed May 19, 2019. https://www.pwc.co.uk/services/sustainability-climate-change/insights/how-ai-future-can-enable-sustainable-future.html

38   "OECD Sustainable Manufacturing Toolkit: Seven Steps to Environmental Excellence." Organisation for Economic Co-operation and Development, 2011. http://www.oecd.org/innovation/green/toolkit/#d.en.192438

39   http://www.oecd.org/innovation/green/toolkit/aboutsustainablemanufacturingandthetoolkit.htm

ENDNOTES

ACKNOWLEDGEMENTS

I would like to thank the following contributors for providing their insights and perspectives in the development of this book.

Barbara Olagaray Gatto, Guy Berger, Hemant Pathak, Indranil Sircar Jane Broom Davidson, Jack Chen, Jeremy Rollison, Karon Kocher, Lucas Joppa, Marcus Bartley Johns, Mark Lange, Mike Phillips, Nick Tsilas, Owen Larter, Portia Wu, Stephanie Rowland, Steve Guggenheimer, Steve Sweetman, Thomas Roca, Tracy Kennedy

And special thanks to John Galligan, our series editor.

Greg Shaw is Senior Director in the Office of the CEO at Microsoft and co-author of Hit Refresh by Satya Nadella (Harper Collins) and The Ability Hacks. Across a long career at Microsoft, Greg has served in many senior communications and community facing roles, including as a writer for Bill Gates and helping create the company's giving program to provide access to computing and the Internet through public libraries.

Prior to Microsoft, Greg was publisher and CEO of Crosscut.com, an online news and opinion magazine. He also served for nearly a decade as an executive at the Bill & Melinda Gates Foundation, overseeing policy and advocacy for the foundation's U.S. Program.

Greg has been appointed to numerous start-up, government and nonprofit boards, and is asked frequently to be a speaker on news, media and publishing. He has a B.A. in journalism from Northeastern State University in Oklahoma. Greg and his wife have two children.

GREG SHAW